# Advances in Desiccant Dehumidification

Vivekh Prabakaran · Kian Jon Chua

# Advances in Desiccant Dehumidification

## From Fundamentals to Applications

Springer

Vivekh Prabakaran
Department of Mechanical Engineering
National University of Singapore
Singapore, Singapore

Kian Jon Chua
Department of Mechanical Engineering
National University of Singapore
Singapore, Singapore

ISBN 978-3-030-80842-6 ISBN 978-3-030-80843-3 (eBook)
https://doi.org/10.1007/978-3-030-80843-3



This Springer imprint is published by the registered company Springer Nature Switzerland AG
The registered company address is: Gewerbestrasse 11, 6330 Cham, Switzerland

# Preface

In many cosmopolitan cities, particularly those in the tropics, a substantial portion of the energy in air-conditioned buildings is required for air dehumidification and moisture control. Accordingly, the process of air dehumidification has become an important function alongside the reduction of sensible load. While heat exchangers are conveniently used to handle the sensible load, research on energy-efficient latent load reduction remains an imperative topic. The conventional vapor compression system employs condensation dehumidification to reduce and regulate latent load. The prominent disadvantage is that the efficiency, measured as Coefficient of Performance (COP), is low. To condense moisture from air, the air has to be cooled below its dew point temperature. The evaporator has to operate at a much lower temperature than the air dew point temperature for moisture condensation to take place, as a result, the energy onerous is on the compressor to consume more energy to facilitate a higher thermal lift between condenser and evaporator.

Besides dew-point condensation, desiccants (solid and liquid), are the next best options to control and regulate the moisture in the air. In desiccant dehumidification process, moist air is dehumidified by the principles of adsorption. There is a direct contact between the humid air and the desiccant. The desiccant adsorbs moisture on its surface and reduces the latent load in the air. However, the adsorption process cannot continue infinitely because of the limited adsorption capacity of the desiccants. The adsorbed water molecules have to be removed from them; this process of regeneration of desiccants is often conducted via the heat energy supplied from industrial waste heat or solar energy. Further, commercially available desiccants, namely, silica-gel and zeolite, though widely used in different industries, are not considered pragmatic as far as air conditioning application is concerned because they require high-quality heat to provide sufficient regeneration temperature often above 100 degree Celsius. In addition, these commercial desiccants often have limited water adsorption capacities; requiring them to regenerate within a relatively short period of time after continuous moisture intake. To address these challenges, the needs to explore and synthesize new desiccating materials coupled with innovative design configurations become more exacting than ever. The content of this book is designed and structured as various road maps ushering readers to the recent technical developments in desiccant science and technology that closely explore unique solutions to provide energy-efficient dehumidification and moisture content.

The proposed book covers the state-of-the-art solid desiccant research and technologies with improved sustainability and energy efficiency objectives. It includes a comprehensive review of the existing desiccants, emphasizes the needs to develop advanced energy-efficient desiccants, different material characterization techniques, and explores exciting frontiers in the technology's practical applications in cooling, water harvesting, and heat pumping systems. The materials have been judiciously selected to convey to readers new interesting perspectives on recent technological developments pertaining to air dehumidification and moisture control. Each chapter further endeavors to provide the readers with the tools necessary to perform similar studies for other thermal systems or processes involving the transfer of heat and mass during different stages of dehumidification and moisture control, i.e., air conditioning. Some fundamental works on desiccant coated heat exchanger transfer processes are presented to illustrate to readers the direct link between fundamental thermal science and practical applications. Henceforth, the reader can leverage on his/her newly acquired knowledge and understanding to initiate better interaction and dialogue between researchers and engineering practitioners. It is worthy to note that, although this book has been divided into chapters with certain air dehumidification themes, the presentation in each chapter does not necessarily fit into neat silos; there are times when an overlapping of information exists.

The introductory chapter presents an overview of the progressive development of solid desiccant dehumidifiers, their impacts on energy efficiency, carbon dioxide, and greenhouse emissions. Several recently developed new generation of ceramic and inorganic desiccants and performance comparison is presented in Chap. 2. This chapter further documents the important issue of binder material selections for different desiccants followed by some of the key limitations of these desiccants in terms of material and engineering design challenges including limitations of integrating desiccants with clean energy sources. Chapter 3 provides an extensive coverage on the latest development on desiccant materials. This chapter primarily covers the next generation of high-performing desiccant materials, namely, superabsorbent polymer-based composite desiccants, and Metal-Organic Frameworks (MOFs). Details on various material characterization techniques and key factors that impact desiccant synthesis processes are judiciously documented. Chapter 4 presents recent works conducted on the development of desiccant-coated heat exchangers specifically for the purpose of air conditioning. The last chapter, Chap. 5, will be of great interest to industrial engineers who are presenting considering or exploring the potential of state-of-the-art desiccants for energy-efficient applications, for example, desiccant coated heat pumps, energy storage, or even atmospheric water harvesting. It further incorporates several case studies on large-scale desiccant air conditioning technologies when applied to the healthcare, electronics manufacturing, and building sectors.

The judiciously selected content of this book is intended to serve as a one-stop archive of known practical solutions that solid desiccant dehumidifiers can offer. The coverage of some new fundamental developments in MOFs and solid desiccant

# Advances in Desiccant Dehumidification

Vivekh Prabakaran · Kian Jon Chua

# Advances in Desiccant Dehumidification

## From Fundamentals to Applications

Springer

Vivekh Prabakaran
Department of Mechanical Engineering
National University of Singapore
Singapore, Singapore

Kian Jon Chua
Department of Mechanical Engineering
National University of Singapore
Singapore, Singapore

ISBN 978-3-030-80842-6 ISBN 978-3-030-80843-3 (eBook)
https://doi.org/10.1007/978-3-030-80843-3



This Springer imprint is published by the registered company Springer Nature Switzerland AG
The registered company address is: Gewerbestrasse 11, 6330 Cham, Switzerland

# Preface

In many cosmopolitan cities, particularly those in the tropics, a substantial portion of the energy in air-conditioned buildings is required for air dehumidification and moisture control. Accordingly, the process of air dehumidification has become an important function alongside the reduction of sensible load. While heat exchangers are conveniently used to handle the sensible load, research on energy-efficient latent load reduction remains an imperative topic. The conventional vapor compression system employs condensation dehumidification to reduce and regulate latent load. The prominent disadvantage is that the efficiency, measured as Coefficient of Performance (COP), is low. To condense moisture from air, the air has to be cooled below its dew point temperature. The evaporator has to operate at a much lower temperature than the air dew point temperature for moisture condensation to take place, as a result, the energy onerous is on the compressor to consume more energy to facilitate a higher thermal lift between condenser and evaporator.

Besides dew-point condensation, desiccants (solid and liquid), are the next best options to control and regulate the moisture in the air. In desiccant dehumidification process, moist air is dehumidified by the principles of adsorption. There is a direct contact between the humid air and the desiccant. The desiccant adsorbs moisture on its surface and reduces the latent load in the air. However, the adsorption process cannot continue infinitely because of the limited adsorption capacity of the desiccants. The adsorbed water molecules have to be removed from them; this process of regeneration of desiccants is often conducted via the heat energy supplied from industrial waste heat or solar energy. Further, commercially available desiccants, namely, silica-gel and zeolite, though widely used in different industries, are not considered pragmatic as far as air conditioning application is concerned because they require high-quality heat to provide sufficient regeneration temperature often above 100 degree Celsius. In addition, these commercial desiccants often have limited water adsorption capacities; requiring them to regenerate within a relatively short period of time after continuous moisture intake. To address these challenges, the needs to explore and synthesize new desiccating materials coupled with innovative design configurations become more exacting than ever. The content of this book is designed and structured as various road maps ushering readers to the recent technical developments in desiccant science and technology that closely explore unique solutions to provide energy-efficient dehumidification and moisture content.

The proposed book covers the state-of-the-art solid desiccant research and technologies with improved sustainability and energy efficiency objectives. It includes a comprehensive review of the existing desiccants, emphasizes the needs to develop advanced energy-efficient desiccants, different material characterization techniques, and explores exciting frontiers in the technology's practical applications in cooling, water harvesting, and heat pumping systems. The materials have been judiciously selected to convey to readers new interesting perspectives on recent technological developments pertaining to air dehumidification and moisture control. Each chapter further endeavors to provide the readers with the tools necessary to perform similar studies for other thermal systems or processes involving the transfer of heat and mass during different stages of dehumidification and moisture control, i.e., air conditioning. Some fundamental works on desiccant coated heat exchanger transfer processes are presented to illustrate to readers the direct link between fundamental thermal science and practical applications. Henceforth, the reader can leverage on his/her newly acquired knowledge and understanding to initiate better interaction and dialogue between researchers and engineering practitioners. It is worthy to note that, although this book has been divided into chapters with certain air dehumidification themes, the presentation in each chapter does not necessarily fit into neat silos; there are times when an overlapping of information exists.

The introductory chapter presents an overview of the progressive development of solid desiccant dehumidifiers, their impacts on energy efficiency, carbon dioxide, and greenhouse emissions. Several recently developed new generation of ceramic and inorganic desiccants and performance comparison is presented in Chap. 2. This chapter further documents the important issue of binder material selections for different desiccants followed by some of the key limitations of these desiccants in terms of material and engineering design challenges including limitations of integrating desiccants with clean energy sources. Chapter 3 provides an extensive coverage on the latest development on desiccant materials. This chapter primarily covers the next generation of high-performing desiccant materials, namely, superabsorbent polymer-based composite desiccants, and Metal-Organic Frameworks (MOFs). Details on various material characterization techniques and key factors that impact desiccant synthesis processes are judiciously documented. Chapter 4 presents recent works conducted on the development of desiccant-coated heat exchangers specifically for the purpose of air conditioning. The last chapter, Chap. 5, will be of great interest to industrial engineers who are presenting considering or exploring the potential of state-of-the-art desiccants for energy-efficient applications, for example, desiccant coated heat pumps, energy storage, or even atmospheric water harvesting. It further incorporates several case studies on large-scale desiccant air conditioning technologies when applied to the healthcare, electronics manufacturing, and building sectors.

The judiciously selected content of this book is intended to serve as a one-stop archive of known practical solutions that solid desiccant dehumidifiers can offer. The coverage of some new fundamental developments in MOFs and solid desiccant dehumidification is often limited in many review papers. The technical content of each chapter will strive to strike a balance between fundamentals and applications to render readers both depth and wide. Due to the broadness of recent scientific developments on air conditioning, the selection of the materials and their balance have been a most difficult task. Pertinent materials have been selected from literature and our published works. These are judiciously put together in an easily digestible format. Credits should belong to the original sources. Lastly, we would like to add that the technical content presented in this book has all been done in the spirit of contributing to the knowledge pool of existing resources on air dehumidification and moisture control—a subject though mature yet still has spacious research room to pursue and innovate.

In the spirit of expressing gratitude, the authors like to extend their heartfelt thanks and appreciation to some team members who have assisted and contributed to the documentation of some of the technical content presented in the various chapters. Some of these people include research staff and ex-Ph.D. students, namely, Md Raisul Islam, Cui Xin, Lin Jie, Bui Duc Thuan, M Kum Ja, Avishek Karmakar, Kyaw Minn Htun, Oh Seung Jin, and other graduate students who have, at various times, worked and contributed immensely during their residence in our laboratories.

Singapore

Vivekh Prabakaran
Kian Jon Chua

# Contents

# About the Authors

**Vivekh Prabakaran** Research Fellow, Department of Mechanical Engineering, National University of Singapore.

Research interests: Dehumidification; New materials for energy research; Air-conditioning; Optimization; Desalination; Solar energy.

Dr. Vivekh Prabakaran currently works as a Postdoctoral Research Fellow with the Department of Mechanical Engineering, National University of Singapore (NUS). He completed his undergraduate coursework from BITS Pilani, India (2015) and worked as a research assistant at the Heat Transfer Laboratory of the Indian Institute of Science. Being the recipient of the prestigious President's Graduate Fellowship, he received his Ph.D. in Mechanical Engineering from NUS in 2020 under Prof. Kian Jon Chua's supervision. During his doctoral training, he synthesized next-generation advanced desiccant materials and conducted key fundamental experimental and theoretical investigations on heat and mass transport. His Ph.D. findings have been published in over 7 peer-reviewed international journals. Vivekh's specific research focus areas include dehumidification, cooling, heat pumping, desalination, solar energy, multi-criteria decision analysis, and advanced materials for energy research.

**Kian Jon Chua** Associate Professor, Department of Mechanical Engineering; National University of Singapore.

Expertise: Air-conditioning, dehumidification, process heating, refrigeration, district cooling, co-generation/tri-generation, thermal water desalination.

Dr. Chua Kian Jon is currently an Associate Professor with the Department of Mechanical Engineering, National University of Singapore. He has been conducting research on air-conditioning, dehumidification, and heat pump systems since 1997. He has conducted both modelling and experimental works for specific building HVAC applications. These include dehumidification/dehydration, cooling, heat pumping, compact heat exchangers and refined temperature/humidity control. He is highly skilled in designing; fabricating; commissioning and testing many experimental test-rigs for heating, cooling and humidity control for both small and large scale applications. He has more than 200 international peer-reviewed journal publications, 6 book chapters and one monograph on advances in air conditioning (https://www.springer.com/gp/book/9789811584763). His works has garnered more than 10,000 over citations. Further, he owns more than 10 patents related to several innovative cooling and dehumidification systems. He is the Principal Investigator of several multi-million competitive research grants. Additionally, he has been awarded multiple local, regional, and international awards for his breakthrough research endeavors.

# Chapter 1
# Progressive Development of Solid Desiccant Dehumidification Technology

**Abstract** Population increase, rapid industrialization, economic growth, and increased demand for thermal comfort have exponentially raised buildings' air-conditioning requirements. Over 90% of the current air-conditioning market is dominated by mechanical vapor compression (MVC) technology, which is essentially a coupled condensation dehumidification-cooling method. Its energy efficiency is low due to significant overcooling and reheating of the process air. In addition, its high electricity consumption contributes to substantial $CO_2$-equivalent emissions and significantly impacts climate change. If the cooling demand continues to grow at the present rate, the MVC cooling technology will alone contribute to almost 0.5 °C rise in global temperatures by 2100s. Therefore, energy-efficient alternative cooling technologies are essential to address the sharp rise in building energy. Scientists worldwide are evolving new cooling technologies and have categorized them into solid-state, electrically driven mechanical, and thermally driven alternatives. Among the 20 alternatives that exist in different stages of research, prototyping, and commercialization, desiccant dehumidification technology decouples latent dehumidification and sensible cooling and has demonstrated to be an excellent solution to promote the energy efficiency of the conventional cooling process. This chapter broadly discusses the future of cooling, highlights key sustainable alternative technologies, and comprehensively reviews the merits of solid desiccant dehumidification technologies.

**Keywords** Air-conditioning · Vapor compression · Energy efficiency · Novel cooling technology · Desiccant dehumidification

## 1.1 Introduction

The global energy demand is projected to increase by 50% from 500 quadrillion British thermal units (Quads) in the 2010s to 850 Quads in 2050s [1]. This rise is attributed to numerous technological advancements, socio-economic, and cultural factors such as population increase, economic growth, and urbanization. Buildings, industry, and transportation sectors are the major energy consumers, with buildings alone accounting for approximately 40% of the energy market [2]. Figure 1.1a shows

 
V. Prabakaran and K. J. Chua, *Advances in Desiccant Dehumidification*,
https://doi.org/10.1007/978-3-030-80843-3_1

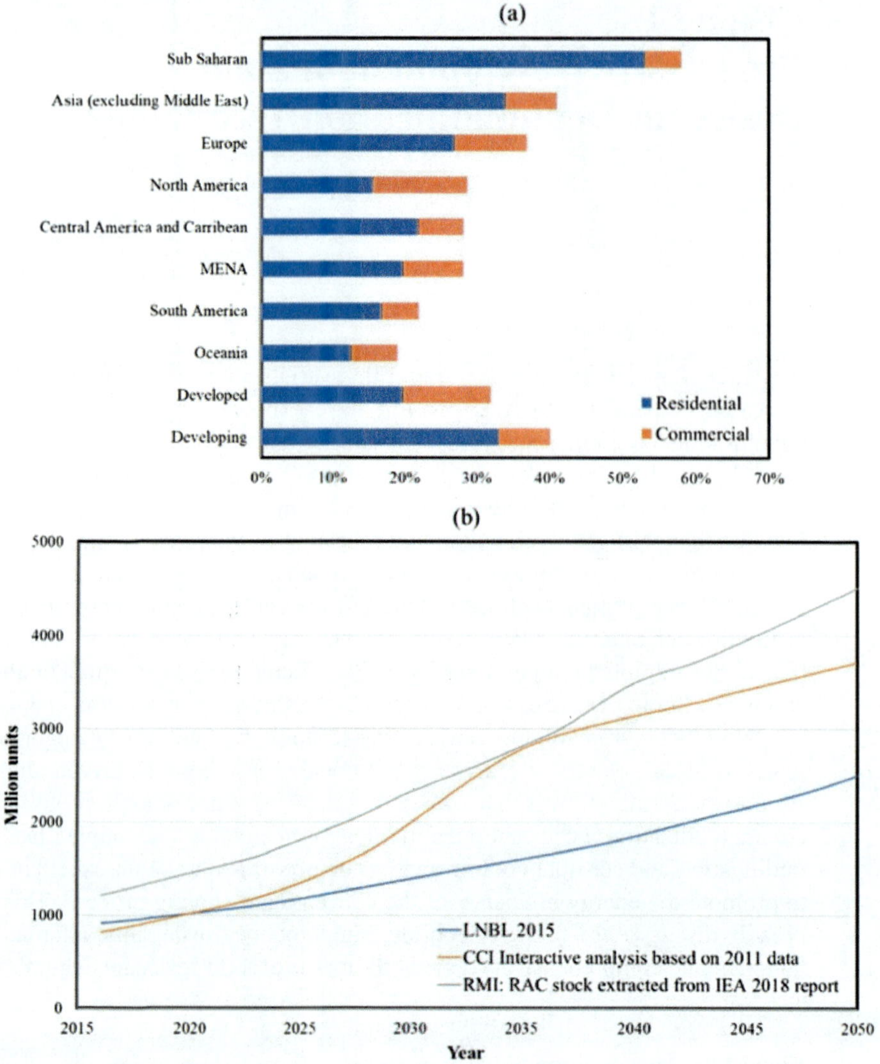


**Fig. 1.1** **a** Energy consumptions in buildings across different regions of the world [3] and **b** projected demand of residential air-conditioning from 2016 to 2050 [4]

that the residential sector utilizes around 70–80% of the building energy consumption, and the air-conditioning requirement constituting up to 50% of the residential building energy [3].

Rising environmental temperature, pollution, and ever-expanding urban islands in many regions make the air-conditioning process an immediate priority to achieve general well-being. Thus, the energy demand for thermal comfort in the residential sector has already tripled in the last 25 years. It continues to grow at an inexorable pace compared to any other end-use (Fig. 1.1b). Due to the many energy-related emissions, the cooling process accounts for a significant risk to global climate, and the residential sector alone is expected to contribute to around 0.5 °C increase in global temperatures by 2100 [4].

## 1.2 Present State of the Air-Conditioning Process

The mechanical vapor-compression (MVC) technology, occupying over 90% of the existing air-conditioning industry market, is adopted due to its smaller size and installation flexibility. The MVC air-conditioners comprise four components: compressor, condenser, expansion valve, and evaporator (Fig. 1.2). The refrigerant first enters the compressor as a low-pressure vapor and discharges as a high-pressure high-temperature vapor. Next, it passes through a condenser, where the refrigerant releases heat with the surroundings and condenses into a high-pressure liquid. Then the refrigerant undergoes an isenthalpic expansion process in the expansion valve, and the resulting low-pressure liquid-vapor refrigerant mixture evaporates by capturing heat from the desired cooling space.

In residential air-conditioners, the outdoor air is directly supplied to the evaporator coil whereas, in the industrial/commercial air-conditioners, the evaporator first cools water to produce chilled water, and the chilled water eventually cools down

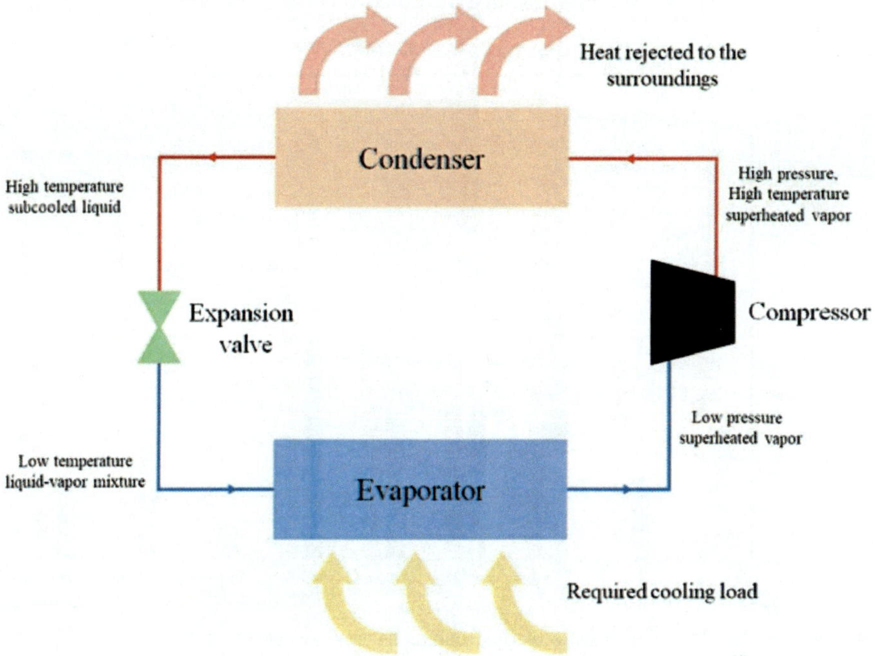


**Fig. 1.2** Working principle of a typical MVC air-conditioning system

the incoming air in an air-handling unit (AHU). The evaporator coil in both cases is maintained much lower than the air's dew point temperature. This is done to reduce both latent and sensible cooling loads in the air simultaneously [5]. The air needs to be overcooled for condensing moisture, and an additional reheating step is often necessary to provide air at an acceptable thermal comfort range of 22–24 °C and 50–65% relative humidity (RH), specified by the American Society of Heating, Refrigerating, and Air-conditioning Engineers (ASHRAE) [6].

Although the MVC systems were first developed more than 100 years ago [7], the compressors have barely reached 14% of their theoretical maximum efficiency, as defined by the Carnot cycle. This results in a large amount of electricity requirement, and with the demand for air-conditioners rising, the corresponding emissions are also increasing equitably. Figure 1.3 illustrates the share of $CO_2$ emissions from residential air-conditioners as a percent of the total primary energy consumed across different regions. The rising air-conditioning demand would result in higher emission levels, which would in turn contribute to the exacerbation of already increasing global average temperature.

At present, the residential air-conditioners generate nearly 700 million metric tons of carbon dioxide-equivalent ($CO_2e$) emissions every year. There are two types of emissions associated with their operation: direct emissions due to chemical-based refrigerants such as hydrofluorocarbons (HFCs) and hydrochlorofluorocarbons (HCFCs) and indirect emissions from extensive electricity production [8]. Approximately 70% of this total is due to indirect emissions, and the rest is accounted by direct emissions [9]. Macroscopically, the emissions due to electricity consumption are the

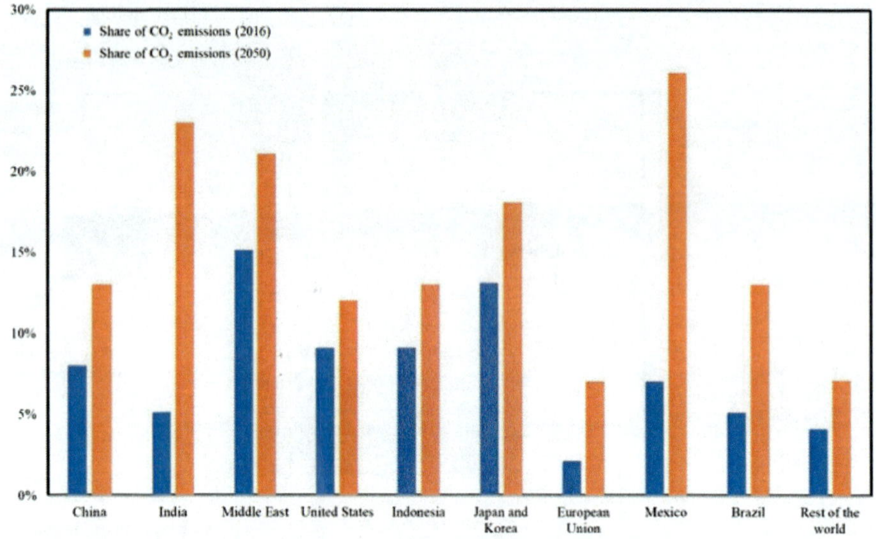


**Fig. 1.3** Percentage share of $CO_2$ emissions from residential air-conditioners across different regions of the world [4]

most significant contributor; however, it is important to appreciate the fact that refrigerants have a disproportionately high global warming impact relative to their mass. Therefore, markedly reducing direct emissions offers a vital path to control the overall $CO_2e$ emissions. In theory, if the refrigerants used in the existing air-conditioners are replaced with near-zero global warming potential (GWP) refrigerants, the annual global emissions could be slashed by almost 20% [10]. With a strong commitment to two recent international agreements—the Paris agreement (December 2015) and the Kigali amendment (2016) to the Montreal Protocol, a significant reduction of direct $CO_2$ emissions has already been achieved. Chlorofluorocarbons (CFCs), which depleted the ozone layer, have been phased out, and HCFCs are expected to be phased out entirely by 2030. Unlike these traditional refrigerants which are saturated organic compounds, unsaturated organic compounds such as hydrofluoroolefins (HFOs) with low GWP are being developed [11]. Besides replacing the refrigerants, it is worth emphasizing that even greater reductions in emissions can be achieved by improving the air-conditioning process efficiency and by minimizing electricity consumption.

## 1.3 Future of Air-Conditioning

The air-conditioning industry has progressively improved its energy efficiency by introducing components such as variable-speed drives, highly efficient fans/motors, and electronic expansion valves, developing advanced control/operation strategies, and designing improved heat exchangers. However, the improvement in coefficient of performance (COP), defined as the ratio of cooling load to electricity consumption, has largely been incremental. This marginal increase is attributed to the inherent limitation of simultaneous handling of dehumidification and cooling processes.

### *1.3.1 Alternative Cooling Solutions*

To build sustainable air-conditioners for the future, thermal scientists worldwide have developed more than 20 different alternatives, and these out-of-the-box solutions can be classified into three categories: electrically powered solid-state devices, electro-mechanical systems, and thermally driven cooling technologies [10] (Fig. 1.4). The most prevalent solid-state cooling systems include magnetocaloric and thermoelectric cooling technologies. These systems are built on advanced paramagnetic/semiconductor materials that produce a heat flux when exposed to changing magnetic/electrical fields (Fig. 1.5). Although these novel substitutes show promising results for improvements in energy efficiency, they are either in their early stages of development or involve higher cost/complexities for practical applications.

The most commonly available electro-mechanical cooling systems are direct evaporative coolers. In this technology, the air is cooled by utilizing the latent heat of vaporization of water. In systems where water is directly evaporated into the supply

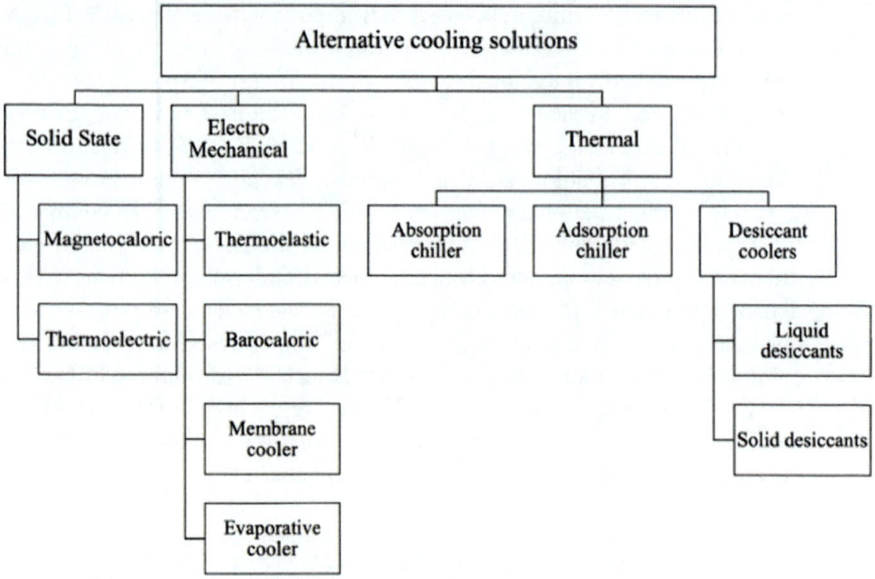


**Fig. 1.4** Classification of alternative cooling solutions

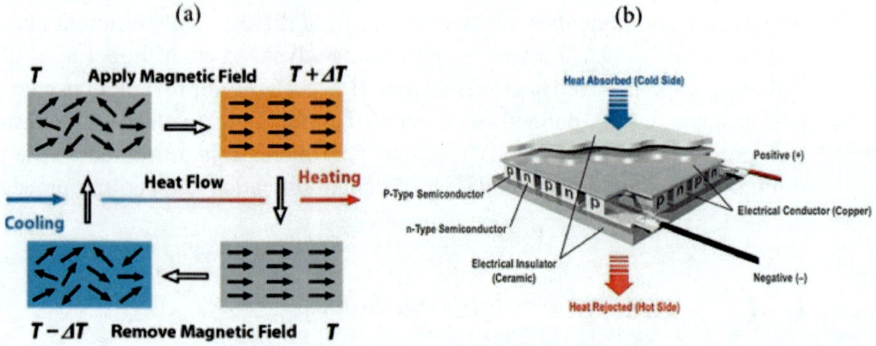


**Fig. 1.5** A schematic representation of **a** magnetocaloric [12] and **b** thermoelectric cooling systems [13]

airstream (Fig. 1.6a), the air temperature drops, and the humidity increases, thus, retaining a constant enthalpy value. These coolers are easy to design, cost-effective, and are useful in arid regions. However, they are not suitable for humid regions since high humidity level in the air overheats the human body, promotes bacterial/fungal growth, and causes respiratory infections. In contrast, in indirect evaporative coolers (Fig. 1.6b), the water cools down an additional airstream passing through a heat exchanger, and the supply air is then cooled without any change in its humidity. Although the indirect ones provide cooler and drier air than their direct counterparts, they are complex to manufacture and usually require a pre-dehumidification step to

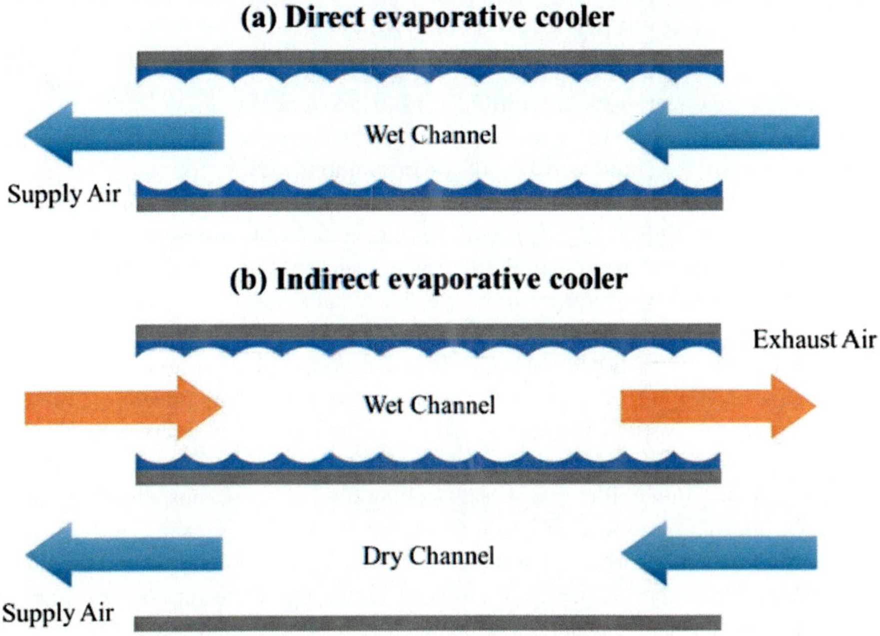


**Fig. 1.6** A schematic of **a** direct and **b** indirect evaporative coolers

operate at the highest cooling effectiveness [14]. In addition, membrane dehumidifiers achieve isothermal dehumidification by transferring moisture from the humid air across a semi-permeable membrane. They are driven by vacuum pumps, which provide transmembrane pressure difference that acts as a driving force for moisture removal [15]. While membrane dehumidifiers are expected to offer almost 50% energy savings than conventional cooling technologies, their key drawback is the appropriate sizing, bulkiness, and noise generated by the vacuum pump [16].

Analogous to solid-state cooling systems, thermoelastic or elastocaloric technology generates a heat flux when an elastic material is stretched and released periodically. The biggest hurdle that prevents this technology from being commercialized is the lack of an efficient method to apply large compression loads in a small footprint with full recovery of the unloading energy. In contrast, barocaloric coolers produce adiabatic temperature changes when external pressure is applied and released repeatedly. The cooling effect is produced due to the changes in the molecular orientation of the solid organic plastic crystals or shape memory alloys. The barocaloric coolers demonstrate high theoretical potential, yet comprehensive efforts to commercialize them have not been carried out [17]. Therefore, the realistic application potential of these caloric cooling solutions require comprehensive fundamental research in material sciences, thermodynamically efficient cycle design, multiple stages of prototype development, and rigorous testing [18].

### 1.3.2 Thermally Driven Alternatives

Absorption chillers are different from conventional MVC chillers as they are compressor-free and use low-grade thermal sources as energy input instead of electricity. A fluid mixture of water-ammonia ($NH_3$) or water-lithium bromide (LiBr) is employed to replace HCFC refrigerants. These chillers compress the refrigerant ($NH_3$/LiBr) by dissolving the vapor in an absorbent (water) and pumping the resulting mixture to a high-pressure location. Figure 1.7a describes the working principle of an absorption chiller. Its disadvantages emerge from the liquid mixture's corrosive nature, bulkiness of the components involved, large footprint, and energy efficiency [19]. Since the COP of absorption chillers is around 1 (approximately 3–4 times lower than the MVC chillers), they are economically viable only when the electricity price is lower than the expense paid for generating heat or a source of waste heat is readily available.

Adsorption chiller employs a working fluid and a porous adsorbent and operates through a reversible adsorption/desorption cycle. As shown in Fig. 1.7b, the working fluid is first adsorbed (at surface level) by the porous material and then released by supplying thermal energy. It employs highly durable, non-corrosive, and environmentally friendly adsorbents, operates at low heat source temperatures compared to absorption chillers, and has maintenance-free operations. Despite the benefits, the adsorption chiller's fundamental drawbacks are its large footprint [22], more extended payback periods, lower adsorption efficiency, and higher irreversibility [23].

A pure thermal approach to enhancing air-conditioning energy efficiency is accomplished by decoupling the simultaneous handling of sensible and latent cooling loads. This is achieved by using desiccant dehumidifiers where the moist air is dehumidified by adsorption/absorption (together referred to as sorption) phenomenon. Desiccant dehumidifiers are highly suitable to function as a standalone unit for reducing humidity, accompany conventional air-conditioners, or can be coupled with other non-conventional cooling systems such as evaporative coolers and membrane dehumidifiers.

The desiccating material, capable of readily capturing moisture from the air, reduces the air's latent load upon its direct contact with the humid air. It cannot sorb moisture indeterminately because of its fixed water capturing capacity. The moisture carried by the desiccant must be removed by supplying heat from waste heat/geothermal/solar collectors [24]. Desiccant dehumidifiers are primarily classified into liquid desiccant and solid desiccant dehumidifiers based on the physical state of the desiccant materials employed.

Liquid desiccant systems consist of an absorber, a regenerator, and a series of heat exchangers. In the absorber, a concentrated liquid desiccant solution, made up of materials such as lithium chloride (LiCl), calcium chloride ($CaCl_2$), LiBr, sodium chloride (NaCl), and tri-ethylene glycol (TEG), absorbs moisture from air due to the vapor pressure difference between the air and desiccant solution surface. As the desiccant solution captures moisture from air, it turns into a weak solution and is

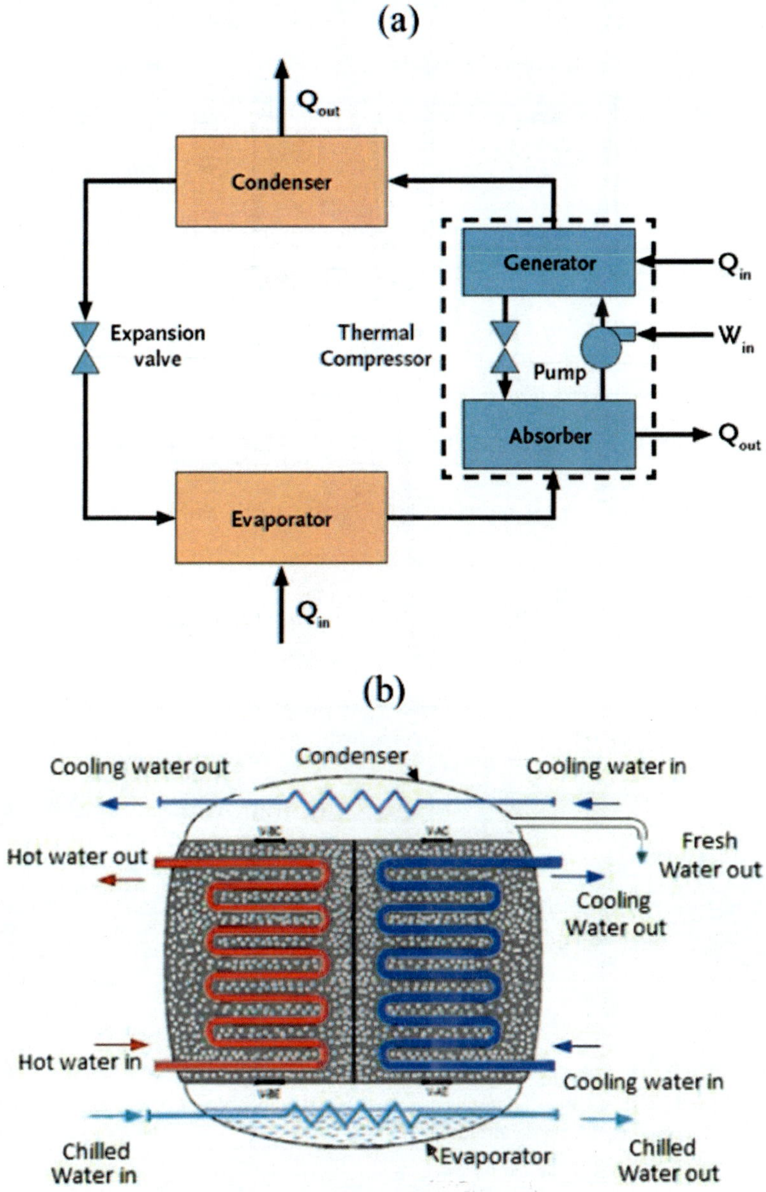


Fig. 1.7 Schematic representation of **a** an absorption chiller [20] and **b** an adsorption chiller [21]

made to pass through the regenerator. Thermal energy is then supplied to evaporate the water absorbed and reactivate the desiccant solution [25]. Figure 1.8a illustrates the working principle of liquid desiccant dehumidifiers.

In solid desiccant dehumidifiers, a stream of air is made to pass through a desiccant matrix, which produces dry but heated air. The most commonly employed solid

(a)

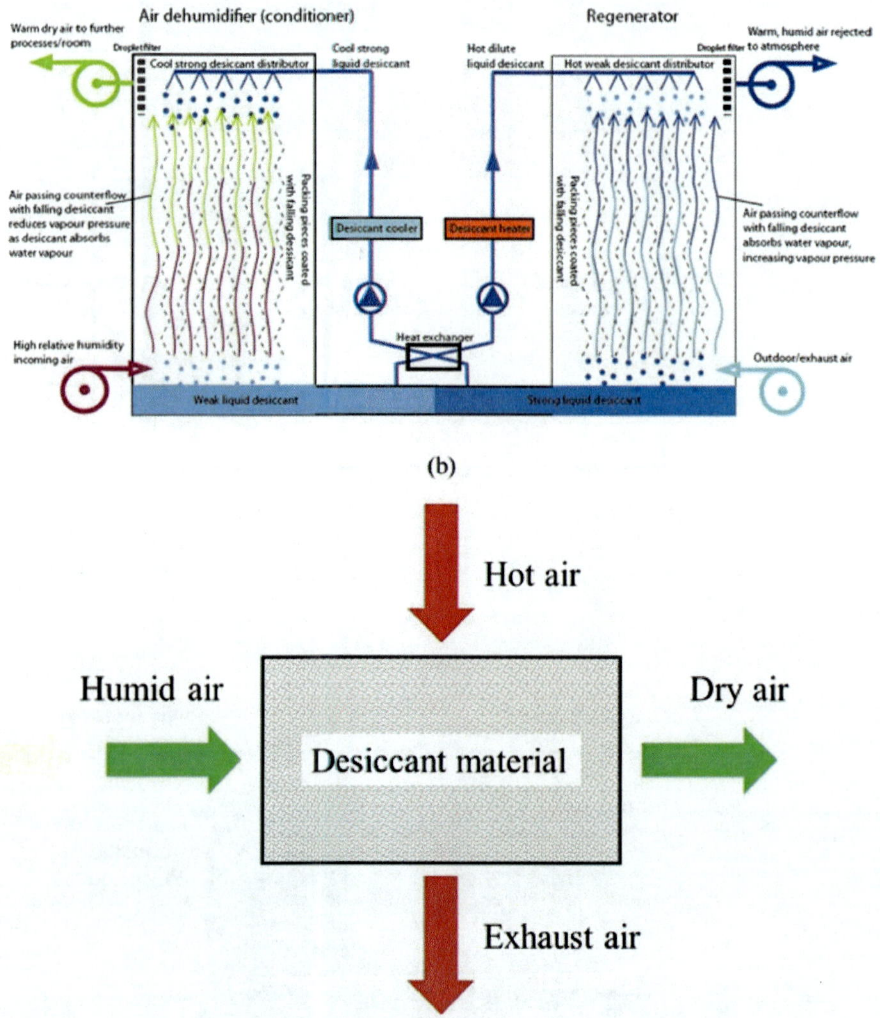


**Fig. 1.8** A schematic illustrating the working principle of **a** a liquid desiccant dehumidifier [27] and **b** a solid desiccant dehumidifier [28]

desiccants are silica gel, zeolites, alumina, and salt composites of these desiccants. Similar to liquid desiccants, the driving force for moisture removal from the air is the vapor pressure difference between the desiccant surface and the process air [26]. The moisture sorption takes place until the desiccant layer is saturated, and heat is supplied for desiccant's regeneration (Fig. 1.8b). A detailed comparison between liquid and solid desiccant dehumidifiers is listed in Table 1.1. These inherent advantages have placed the solid desiccant dehumidification systems as the potential successors to MVC systems.

**Table 1.1** Comparison between liquid and solid desiccant dehumidifiers

| Characteristics | Liquid desiccant dehumidifiers | Solid desiccant dehumidifiers |
|---|---|---|
| Corrosion | Salt solutions are used as liquid desiccants. These salt solutions are highly corrosive to the metal components of air-conditioning systems | Solid desiccants such as silica gel, zeolites, activated carbons, and superabsorbent polymers have no corrosion issue on the metal components of the air-conditioning systems |
| Crystallization | Liquid desiccants are capable of crystallizing when their concentration exceeds 50% and at low temperatures. This would cause the system to stop working | The solid desiccants do not pose any crystallization problems |
| Carry-over | The carry-over of liquid desiccant droplets into the leaving airstream is a major drawback as the air quality is significantly affected | Carry-over effects are not observed in solid desiccant dehumidifiers |
| Reverse dehumidification | Vapor pressure of liquid desiccants is high, resulting in lower driving force of moisture absorption, hence, reversal of air dehumidification to humidification is possible | Vapor pressure of solid desiccants is much lower, resulting in higher driving force for moisture sorption and no reversal of air dehumidification to humidification issue |
| Regeneration temperature | A low regeneration temperature requirement of 40–70 °C | Temperature spanning 80–120 °C is needed for effective regeneration. However, with recent advancements in material science, solid desiccant can be regenerated between 40 and 60 °C |

## 1.4 Solid Desiccant Dehumidification Systems

In this section, several recent advancements in solid desiccant dehumidifiers are presented. On the basis of dehumidifier design and air-flow configuration, solid desiccant dehumidifiers are broadly classified into three types—stationary beds, rotary wheels, and cross-cooled dehumidifiers.

### *1.4.1 Stationary Beds*

#### 1.4.1.1 Packed/Fixed-Bed Systems

Packed/Fixed bed systems are one of the first designs due to their simplicity and ease of fabrication. In this type, a desiccant matrix is tightly packed into a stationary

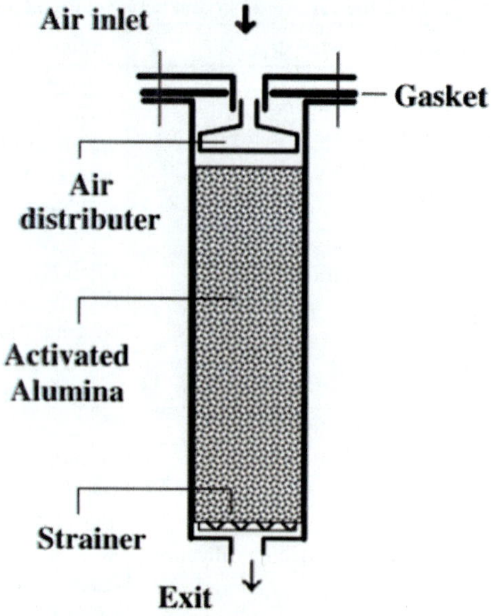


**Fig. 1.9** A schematic of fixed-bed dehumidifier [29]

bed (a heat exchanger or a plate), and the moist air flows through it (Fig. 1.9) [29]. The dry desiccant then readily captures moisture and produces dry air. Since the desiccant matrix has a finite moisture removal capacity, reactivating it for the next dehumidification cycle is necessary. This is achieved by passing a stream of hot air, typically between 80 and 100 °C, through the desiccant matrix. Since the dehumidification process is not continuous using a single bed, at least two beds are necessary to produce a continuous supply of dry air. As a result, the required footprint per dehumidification capacity is high, thereby contributing to the system's major drawback. Further, poor contact between the desiccant and the bed lowers the overall heat transfer coefficient. The exothermic heat released by the desiccant upon capturing moisture remains trapped in the desiccant matrix which tends to negatively impacts the dehumidification capacity of the fixed bed dehumidifiers [30].

#### 1.4.1.2 Fluidized Beds

A fluidized bed is a heterogeneous mixture of a solid (desiccant granules) and a fluid (supply/regeneration air), which exhibits fluid-like physical characteristics. Silica gel, zeolites, and activated alumina are some of the more popularly used materials in fluidized beds, and the latest developments indicate a paradigm shift to employ advanced desiccants such as the Metal-Organic Framework (MOF) developed by the Hong Kong University of Science and Technology (HKUST-1) [31].

In contrast to packed-beds, fluidizing the solid desiccant particles improves the heat and mass transfer rates and results in a 20% improvement in dehumidification performance [29]. The pressure drop due to the tightly packed desiccant matrix in

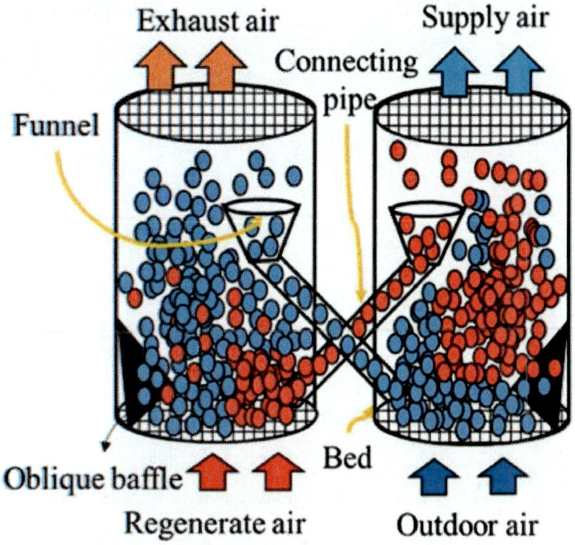


**Fig. 1.10** A schematic of a fluidized bed dehumidifier [33]

the fixed bed system is also reduced. The key disadvantage of this system is due to a potential carryover effect as a result of direct mixing of the airstream with a high volume of desiccant granules. Further, the additional power required to fluidize the particles and maintain a cyclic operation reduces the system's dehumidification energy efficiency. In order to cut down this excess power requirement, funnels are employed (Fig. 1.10) to intercept the falling desiccant particles and transfer them to the other bed [32, 33]. It is also worthy to note that appropriate sealing of the desiccant granules in the bed and air leakage are critical factors that influence fluidized beds' design.

#### 1.4.1.3 Multilayer Fixed-Bed Binder-Free Desiccant Dehumidifiers

Recently, prototypes of multilayer fixed-bed binder-free desiccant dehumidifiers (MFBDDs) have been developed, and one of these prototypes is depicted in Fig. 1.11. In this system, several sheet-type layers of solid desiccants are packed in stainless steel meshes. The sheet-like packing eliminates the need for binders and also reduces the pressure drop in airflow as compared to packed-bed systems. The marked disadvantage of this system is the release of exothermic heat in the airstream causing 8–10 °C temperature rise. Since temperature increase deteriorates the desiccant's moisture capturing capacity, an intercooling step is necessary to address the problem of undesired temperature rise and realize this technology's highest potential [34, 35].

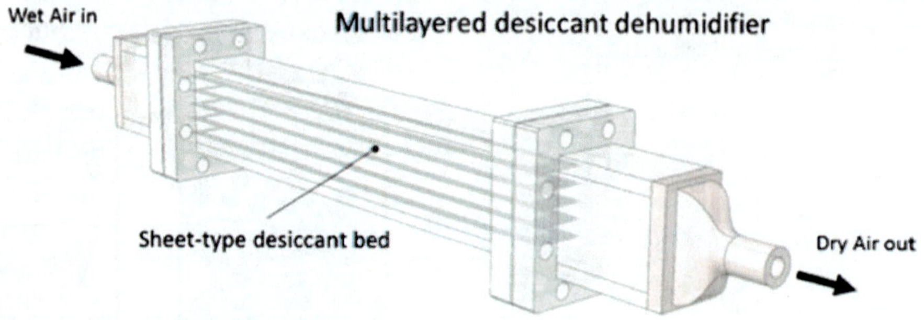


Fig. 1.11 A schematic of multilayer fixed-bed binder-free desiccant dehumidifiers (MFBDDs) [35]

## 1.4.2 Rotary Wheels

Rotary desiccant systems or desiccant wheels are the most prevailing solid desiccant dehumidifiers in the market. In desiccant wheels, solid desiccants are impregnated on a wheel, and often organic/inorganic binders are necessary to achieve uniformity in desiccant impregnation. As illustrated in Fig. 1.12a, two-thirds of the wheel is utilized for air dehumidification, while in the other portion, hot air regenerates the desiccant [36]. The wheel rotates at a constant speed of 6–12 rotations per hour making the

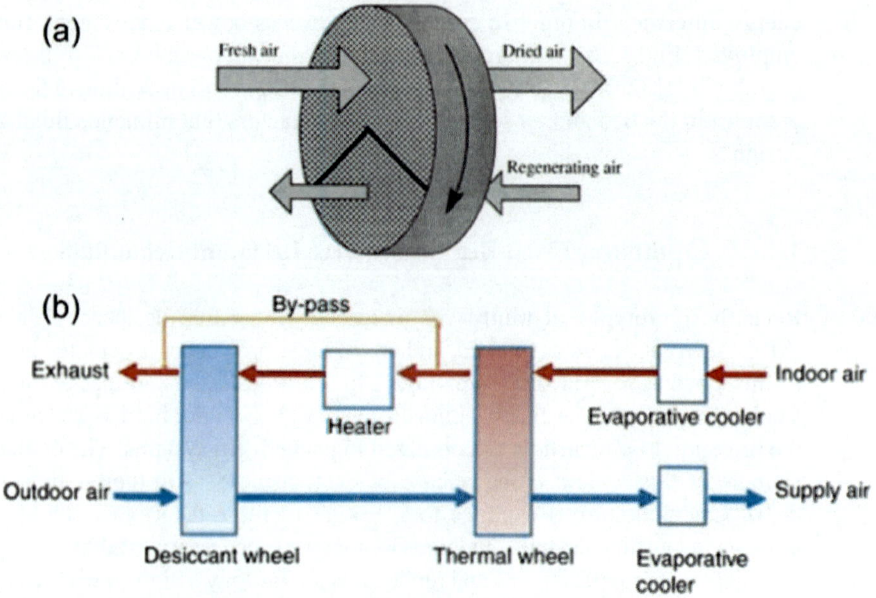


Fig. 1.12 A schematic of **a** a standalone desiccant wheel [36] and **b** a desiccant wheel integrated with an enthalpy wheel and an evaporative cooler [39]

dehumidification process continuous [37]. Consequently, its footprint per dehumidification capacity is much smaller than the stationary-bed systems. Desiccant impregnation improves the contact between the air molecules and the desiccant matrix. This results in higher utilization of the desiccant material and substantial improvements in heat transfer coefficients [38]. However, the exothermic heat released by the desiccant raises the air temperature and evaporative coolers/heat recovery devices are usually installed to remove the excess heat added by the desiccant (Fig. 1.12b). Besides, the complex honeycomb structure of the desiccant wheel results in pressure drops and also leads to potential blockages for airflow (Fig. 1.13).

A new design of a water-cooled desiccant wheel, similar to a shell and tube heat exchanger, has been proposed to reduce the excess heat addition in air as a result of the release of exothermic heat. The outdoor humid air is made to pass through the bottom part of the wheel, where dehumidification takes place at the internal side of the tubes. In the top part, hot air at 50–60 °C facilitates appropriate regeneration. A heat transfer fluid is circulated on the tube's external surface, which assists in removing the released heat of adsorption by the desiccant material during dehumidification. However, due to the design complexities, more often than not, the heat transfer process does not occur efficiently in the innermost layers causing the air temperature

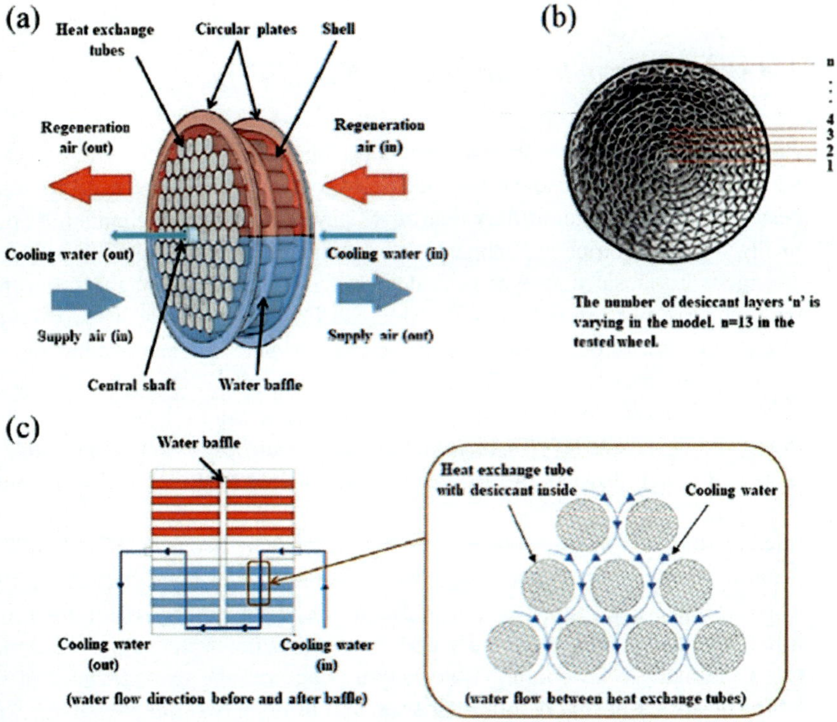


**Fig. 1.13** A schematic of **a** a water-cooled desiccant wheel, **b** inner surface of the tubes showing desiccant impregnation, and **c** the water circulation path [40]

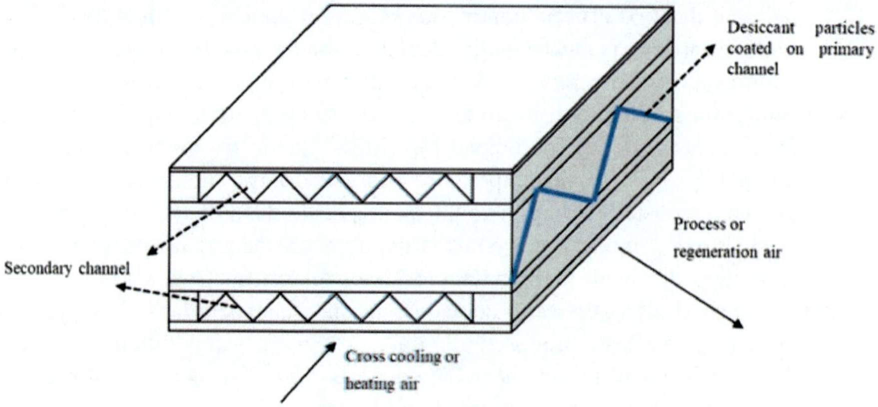


**Fig. 1.14** Schematic of an air-air cross-cooled compact dehumidifier [42]

to rise rapidly. Despite the unfavorable temperature increase, this new configuration achieved 48% higher dehumidification capacity compared to the commonly found desiccant wheels [40, 41].

### *1.4.3 Cross-Cooled Compact Dehumidifiers*

Cross-cooled compact dehumidifiers have been developed to address the drawbacks of stationary-bed and rotary desiccant systems. One of the first crossed-cooled desiccant-coated dehumidifiers is an air-to-air compact heat exchanger. As portrayed in Fig. 1.14, it comprises primary and secondary airflow channels in a crossflow arrangement. Desiccant layer is coated on the primary air channels and the moist outdoor air flows through this channel. Similar to the rotary wheels, binders are often necessary to ensure continuous contact between the heat exchanger channels and desiccant materials. Cooling air is passed through the secondary channel during dehumidification process to capture the exothermic heat released by the desiccant. When the desiccant layer saturates, the cooling airflow is replaced with a stream of hot air which flows in the secondary channels for reactivating the desiccant [42] (Fig. 1.15).

Since the heat transfer rate of water is faster than air, air-to-liquid cross-cooled dehumidifiers have also been built, where the desiccant layer is coated on fin-tube heat exchangers. In the desiccant coated heat exchangers (DCHEs), the process air flows over the fin-side externally and the cooling/hot water is passed through the tubes internally. The cooling water supply captures the exothermic heat released by the desiccant during dehumidification and offers a twofold advantage over other types of dehumidifiers. Firstly, the dehumidification process is largely isothermal, and therefore has negligible effect on the desiccant's moisture capturing ability. The outlet air humidity ratio is also improved by about 2.5 g/kg in DCHEs as compared

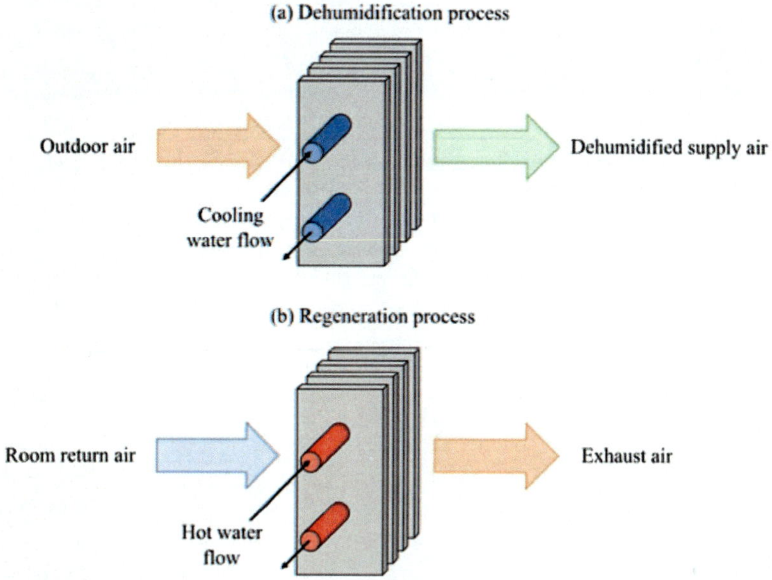


**Fig. 1.15** Schematic representation of **a** dehumidification and **b** regeneration processes in DCHEs [44]

to stationary-beds or rotary wheel dehumidifiers. Secondly, the exothermic heat is not rejected in the supply airstream, and the outlet air temperature is also about 1–3 °C lower than the ambient air temperature [43]. In contrast, the air temperature is usually 5–10 °C higher than the ambient temperature in most stationary beds and rotary wheels. Table 1.2 lists the detailed comparison between the three types of solid desiccant dehumidifiers.

## 1.5 Relevance of Solid Desiccant Dehumidification

With rapid urbanization and higher thermal comfort requirements, the future of desiccant dehumidifiers as a potential substitute for conventional air-conditioning systems is gaining traction. The next-generation dehumidifiers can either be retrofitted with existing air-conditioning equipment or integrated with other heat transfer/exchange devices like heat recovery wheels, solar thermal and photovoltaic collectors, cooling towers, evaporative coolers, and membrane dehumidifiers. Thus, this technology remains vital and will directly impact numerous energy-related technologies such as outdoor coolers, heat pumps, sorption chillers, atmospheric water harvesters, indoor humidity control, and energy storage. Its application will span across a wide range of industries, including but not limited to education, supermarkets, pharmaceutical, electronics, biotechnology, and hospitals.

**Table 1.2** Advantages of cross-cooled dehumidifiers over stationary-beds and rotary wheel dehumidifiers

| Characteristics | Stationary beds | Rotary wheels | Cross cooled dehumidifiers |
|---|---|---|---|
| Moisture removal ability | Low. Temperature increase in the desiccant reduces its moisture removal ability | | High because of simultaneous cooling |
| Total cooling load | The total cooling load remains the same since the exothermic heat is released in the supply airstream. The latent load reduction is converted to an equivalent increase in sensible load | | Cooling water flow removes a significant portion of sorption heat and the total cooling load is reduced |
| Heat transfer efficiency | Low due to higher irreversibility associated with multiple heat transfer resistances | | Internal heating using water flow in the tubes and high thermal conductivity of fins improves the heat transfer efficiency |
| Continuous dehumidification | Requirement of at least two beds | Possible using a single rotary wheel | Requirement of at least two beds |
| Pressure drop | • High for packed beds and<br>• Low for fluidized beds and MFBDDs | Low | Low |
| Desiccant utilization | • Low for packed and fluidized beds<br>• High for MFBDDs since thin sheets of desiccant layer are fixed | High | High |

Due to these noteworthy benefits, the key motivation of this book is to comprehensively document the latest trends in solid desiccant dehumidifiers and evaluate their performance improvement scope. The cooling and thermal performance of desiccant dehumidifiers are highly dependent on the choice of the desiccant material. Thus, a comprehensive review of the physical, chemical, and thermal properties of desiccant materials will be carried out in subsequent chapters. This text will consider pure and composite materials that absorb water vapor due to a combination of several mechanisms notwithstanding adsorption. It will also identify the challenges associated with designing and engineering energy-efficient dehumidifiers based on existing desiccants and establish a need to synthesize new advanced energy-efficient desiccant materials.

Additionally, the fundamental characterization properties that impact dehumidification capacity and thermal efficacy will be analyzed, and critical factors related to desiccant synthesis will also be emphasized. The text will also contain experimental, theoretical, thermodynamic, heat transfer, and economic analyses and evaluate the desiccant dehumidifier's feasibility in energy-related applications. Toward the end,

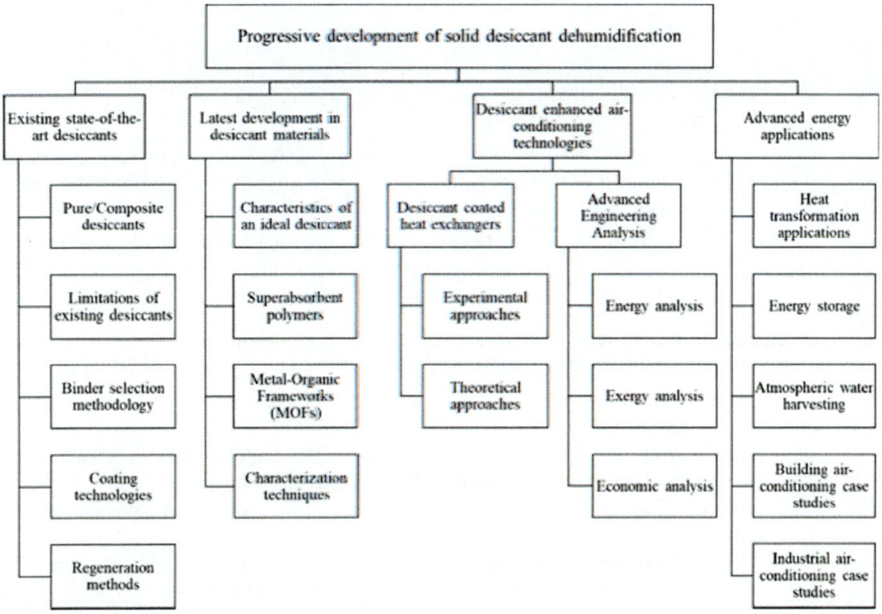


**Fig. 1.16** An outline on the progressive development of solid desiccant dehumidification technology

specific case studies in connection to industrial and building dehumidification will be included. The ensuing chapters of the book will follow the outline as presented in Fig. 1.16.

## 1.6 Conclusions

Rapid urbanization, climate change, and thermal comfort requirements have increased building energy demand sharply. Conventional mechanical vapor compression air conditioners are energy-intensive. They utilize an inefficient simultaneous cooling and dehumidification methodology and significantly raise the global warming potential. Energy-efficient, clean energy-powered air-conditioners are thus necessary to build a sustainable future. Although numerous next-generation cooling technologies have been developed, thermally driven desiccant dehumidifiers show promising potential in improving the air-conditioning process efficiency. The recent developments in solid desiccant dehumidifiers record substantial improvements in dehumidification ability and heat transfer efficiency. As a result of these developments, the new dehumidification technologies directly impact numerous energy-related applications, namely, outdoor coolers, heat pumps, sorption chillers, atmospheric water harvesters, indoor humidity control, and energy storage.

## References


1. Energy Information Administration USEI. Internaltional Energy Outlook 2019. US Energy Inf Adm 2019;September:25–150.
2. Yau YH, Hasbi S. A review of climate change impacts on commercial buildings and their technical services in the tropics. Renew Sustain Energy Rev. 2013;18:430–41. https://doi.org/10.1016/j.rser.2012.10.035.
3. Nejat P, Jomehzadeh F, Taheri MM, Gohari M, Muhd MZ. A global review of energy consumption, $CO_2$ emissions and policy in the residential sector (with an overview of the top ten $CO_2$ emitting countries). Renew Sustain Energy Rev. 2015;43:843–62. https://doi.org/10.1016/j.rser.2014.11.066.
4. Campbell I, Kalanki A, Sachar S. Solving the global cooling challenge, GCP 2018:42.
5. Chua KJ, Chou SK, Yang WM, Yan J. Achieving better energy-efficient air conditioning—a review of technologies and strategies. Appl Energy. 2013;104:87–104. https://doi.org/10.1016/j.apenergy.2012.10.037.
6. Daou K, Wang RZ, Xia ZZ. Desiccant cooling air conditioning: a review. Renew Sustain Energy Rev. 2006;10:55–77. https://doi.org/10.1016/j.rser.2004.09.010.
7. Oppelt D, Papst I, Gloel J, Heubes J, Becker C. Green cooling technologies—market trends in selected refrigeration and air conditioning subsectors 2014.
8. Calm JM. Emissions and environmental impacts from air-conditioning and refrigeration systems. Int J Refrig. 2002;25:293–305. https://doi.org/10.1016/S0140-7007(01)00067-6.
9. Air conditioning is threatening our ability to tackle climate change. Here's what we need to do—Climate Change The New Economy n.d. https://climatechange-theneweconomy.com/air-conditioning-is-threatening-our-ability-to-tackle-climate-change-heres-what-we-need-to-do/. Accessed January 16, 2021.
10. Goetzler W, Zogg R, Jim Young CJ. Energy savings potential and RD & D opportunities for non- vapor-compression HVAC. Energy Effic Renew Energy 2014:3673. https://doi.org/10.2172/1220817.
11. Dai Y, Wang R, Zhang H. Parameter analysis to improve rotary desiccant dehumidification using a mathematical model. Int J Therm Sci. 2001;40:400–8. https://doi.org/10.1016/S1290-0729(01)01224-8.
12. Magnetocaloric Air Conditioning, the Future of AC n.d. https://access-inc.com/the-future-of-air-conditioning/. Accessed January 16, 2021.
13. Thermoelectric cooler solutions for medical applications | Medical Design and Outsourcing n.d. https://www.medicaldesignandoutsourcing.com/thermoelectric-cooler-solutions-for-medical-applications/. Accessed January 16, 2021.
14. Lin J, Thu K, Bui TD, Wang RZ, Ng KC, Chua KJ. Study on dew point evaporative cooling system with counter-flow configuration. Energy Convers Manag. 2016;109:153–65.
15. Bui DT, Nida A, Ng KC, Chua KJ. Water vapor permeation and dehumidification performance of poly(vinyl alcohol)/lithium chloride composite membranes. J Memb Sci. 2016;498:254–62. https://doi.org/10.1016/j.memsci.2015.10.021.
16. Labban O, Chen T, Ghoniem AF, Lienhard JH, Norford LK. Next-generation HVAC: Prospects for and limitations of desiccant and membrane-based dehumidification and cooling. Appl Energy. 2017;200:330–46. https://doi.org/10.1016/j.apenergy.2017.05.051.
17. Zero-GWP cooling using the barocaloric effect—CIBSE Journal n.d. https://www.cibsejournal.com/technical/barocaloric-cooling-a-potential-alternative-to-refrigerant/. Accessed February 10, 2021.
18. Qian S. Development of thermoelastic cooling systems. J Chem Inf Model. 2018;53:1689–99.
19. Srikhirin P, Aphornratana S, Chungpaibulpatana S. A review of absorption refrigeration technologies. Renew Sustain Energy Rev. 2000;5:343–72. https://doi.org/10.1016/S1364-0321(01)00003-X.
20. Module 10: Absorption refrigeration—CIBSE Journal n.d. https://www.cibsejournal.com/cpd/modules/2009-11/. Accessed February 10, 2021.

21. AL-Dadah R, Mahmoud S, Elsayed E, Youssef P, Al-Mousawi F. Metal-organic framework materials for adsorption heat pumps. Energy 2020;190:116356. https://doi.org/10.1016/j.energy.2019.116356.
22. Jagirdar M, Lee PS, Ho GW. Feasibility Study of a Parallel Plate Desiccant Coated Heat and Mass Regenerator for Dehumidification. Energy Procedia. 2017;105:5034–9. https://doi.org/10.1016/j.egypro.2017.03.1011.
23. Demir H, Mobedi M, Ülkü S. A review on adsorption heat pump: Problems and solutions. Renew Sustain Energy Rev. 2008;12:2381–403. https://doi.org/10.1016/j.rser.2007.06.005.
24. Saeed A, Al-Alili A. A review on desiccant coated heat exchangers. Sci Technol Built Environ. 2017;23:136–50. https://doi.org/10.1080/23744731.2016.1226076.
25. Sahlot M, Riffat SB. Desiccant cooling systems: a review. Int J Low-Carbon Technol 2016:ctv032. https://doi.org/10.1093/ijlct/ctv032.
26. Jani DB, Mishra M, Sahoo PK. Solid desiccant air conditioning—a state of the art review. Renew Sustain Energy Rev. 2016;60:1451–69. https://doi.org/10.1016/j.rser.2016.03.031.
27. Module 71: Liquid desiccants for dehumidification in building air conditioning systems—CIBSE Journal n.d. https://www.cibsejournal.com/cpd/modules/2014-12/. Accessed January 9, 2020.
28. Sultan M, El-Sharkawy II, Miyazaki T, Saha BB, Koyama S. An overview of solid desiccant dehumidification and air conditioning systems. Renew Sustain Energy Rev. 2015;46:16–29. https://doi.org/10.1016/j.rser.2015.02.038.
29. Hamed AM, Abd El Rahman WR, El-Eman SH. Experimental study of the transient adsorption/desorption characteristics of silica gel particles in fluidized bed. Energy 2010;35:2468–83. https://doi.org/10.1016/j.energy.2010.02.042.
30. Yanagi H, Ino N. Heat and mass transfer characteristics in consolidated silica gel/water adsorption–cooling system. ASME 1997 Turbo Asia Conf., American Society of Mechanical Engineers; 1997, p. V001T13A009-V001T13A009.
31. Gargiulo V, Raganati F, Ammendola P, Alfe M, Chirone R. HKUST-1 metal organic framework as CO2adsorbent in a sound assisted fluidized bed. Chem Eng Trans. 2015;43:1087–92. https://doi.org/10.3303/CET1543182.
32. Chen CH, Ma SS, Wu PH, Chiang YC, Chen SL. Adsorption and desorption of silica gel circulating fluidized beds for air conditioning systems. Appl Energy. 2015;155:708–18. https://doi.org/10.1016/j.apenergy.2015.06.041.
33. Chiang YC, Chen CH, Chiang YC, Chen SL. Circulating inclined fluidized beds with application for desiccant dehumidification systems. Appl Energy. 2016;175:199–211. https://doi.org/10.1016/j.apenergy.2016.05.009.
34. Hsu WL, Paul S, Shamim JA, Kitaoka K, Daiguji H. Design and performance evaluation of a multilayer fixed-bed binder-free desiccant dehumidifier for hybrid air-conditioning systems: Part II—Theoretical analysis. Int J Heat Mass Transf. 2018;116:1370–8. https://doi.org/10.1016/j.ijheatmasstransfer.2017.09.080.
35. Shamim JA, Hsu WL, Kitaoka K, Paul S, Daiguji H. Design and performance evaluation of a multilayer fixed-bed binder-free desiccant dehumidifier for hybrid air-conditioning systems: Part I—experimental. Int J Heat Mass Transf. 2018;116:1361–9. https://doi.org/10.1016/j.ijheatmasstransfer.2017.09.051.
36. Zhang L-Z. Conjugate Heat and Mass Transfer in Adsorbent Ducts. 2013. https://doi.org/10.1016/b978-0-12-407782-9.00002-2.
37. Narayanan R. Heat-driven cooling technologies. Elsevier Inc.; 2017. https://doi.org/10.1016/B978-0-12-805423-9.00007-7.
38. Ge TS, Li Y, Wang RZ, Dai YJ. A review of the mathematical models for predicting rotary desiccant wheel. Renew Sustain Energy Rev. 2008;12:1485–528. https://doi.org/10.1016/j.rser.2007.01.012.
39. Hundy GF, Trott AR, Welch TC. The Refrigeration Cycle. Refrig. Air Cond. Heat Pumps, Elsevier; 2016, p. 19–39. https://doi.org/10.1016/b978-0-08-100647-4.00002-4.
40. Zhou X, Goldsworthy M, Sproul A. Performance investigation of an internally cooled desiccant wheel. Appl Energy. 2018;224:382–97. https://doi.org/10.1016/j.apenergy.2018.05.011.

41. Shamim JA, Hsu WL, Paul S, Yu L, Daiguji H. A review of solid desiccant dehumidifiers: current status and near-term development goals in the context of net zero energy buildings. Renew Sustain Energy Rev. 2021;137: https://doi.org/10.1016/j.rser.2020.110456.
42. Weixing Y, Yi Z, Xiaoru L, Xiugan Y. Study of a new modified cross-cooled compact solid desiccant dehumidifier. Appl Therm Eng. 2008;28:2257–66. https://doi.org/10.1016/j.applthermaleng.2008.01.006.
43. Ge TS, Dai YJ, Wang RZ, Peng ZZ. Experimental comparison and analysis on silica gel and polymer coated fin-tube heat exchangers. Energy. 2010;35:2893–900. https://doi.org/10.1016/j.energy.2010.03.020.
44. Vivekh P, Bui DT, Wong Y, Kumja M, Chua KJ. Performance evaluation of PVA-LiCl coated heat exchangers for next-generation of energy-efficient dehumidification. Appl Energy 2019;237:733–50.

# Chapter 2
# Current State-of-the-Art in Desiccant Dehumidifiers

**Abstract** Desiccant coated heat exchangers (DCHEs) yield higher dehumidification and thermal efficiency over other solid desiccant dehumidifiers due to their effective removal of the exothermic heat of sorption and improved heat transfer effectiveness. Accordingly, they offer prospective energy and cost savings to several energy-related applications such as heat pumps, chillers, water harvesters, etc. A comprehensive review of the current state-of-the-art in DCHEs is imperative to understand this technology's marked impact and its performing strategies and capabilities. This chapter first introduces the different types of isotherm and hysteresis profiles and specifies the adsorption mechanisms. Then it presents a list of conventional pure/composite desiccants employed and highlights their limitations. Next, the detailed steps involved during its binder material selection are presented, and a comparison is made between the different types of coating techniques. Different regeneration techniques are then described, and the relevance of thermal regeneration vis-à-vis microwave and ultrasonic methods is established. Lastly, the ideal characteristics of a desiccant are listed, which would pave the way for performance-enhancing synthesis of advanced desiccant materials.

**Keywords** Ideal desiccant · Isotherms · Coating techniques · Binders · Regeneration

## 2.1 Introduction

In the introductory chapter, the need for developing alternative cooling solutions was clearly established. Comparisons between electrically driven, electro-mechanical, and thermal alternatives were assessed, and the working principles of different types of solid desiccant dehumidifiers were also provided. It is worthy to note that the dehumidification performance of fixed-bed and rotary wheel dehumidifiers has not reached its maximum potential due to a lack of simultaneous cooling process. The heat of sorption needs to be removed to achieve a more pronounced dehumidification effect. DCHEs have demonstrated better moisture capturing ability and have enhanced heat transfer rates. They can also be coupled with renewable energy sources

 
V. Prabakaran and K. J. Chua, *Advances in Desiccant Dehumidification*,
https://doi.org/10.1007/978-3-030-80843-3_2

and can be readily integrated with numerous energy-related applications. Consequently, an extensive review of the state-of-the-art in DCHEs is quintessential for its practical implementation in wide-ranging applications [1]. On that front, this chapter focuses on the state-of-the-art in desiccants, binder materials, coating techniques, and different types of regeneration methods. The limitations of the conventional desiccant materials and characteristics of ideal desiccant to achieve enhanced dehumidification and cooling performance are also presented.

## 2.2 Water Adsorption Isotherms

Water adsorption isotherms measure the maximum amount of water vapor that a desiccant can adsorb under equilibrium at a particular temperature and relative humidity. The isotherms serve as an essential tool in establishing a desiccant's suitability for different heat transformation applications. Initially, the adsorption isotherms were classified into six types based on the 1985 IUPAC recommendations [2]. However, the recent advancements in material science demanded modifications to the original recommendations. The classifications were updated in 2015, accounting for such developments, and are shown in Fig. 2.1 [3]. The forward direction of the arrow indicates the adsorption process, while the backward arrow direction means the desorption process. It is worthy to note that Types I(a), I(b), II, III, IV (b), and VI are completely reversible during the adsorption and desorption processes whereas hysteresis is observed in other cases.

### *2.2.1 Isotherm Types*

Microporous materials (i.e., pore width not exceeding 2 nm) show strong interactions between the water molecules (the molecules which are adsorbed by the desiccants are generally referred to as adsorbates) and the solid desiccant (a substance that adsorbs adsorbate molecules is known as adsorbent). They are found to have reversible Type I isotherms. The adsorbate molecules fill the adsorbent's pores in the low relative pressure region. This process is entirely reversible [4], predominantly occurs in a single layer, and is referred to as monolayer adsorption. Materials with narrow micropores of width less than ~1 nm show Type I(a) isotherms, and the ones with wider micropores and narrow mesopores (less than ~2.5 nm) record Type I(b) isotherms.

Type II isotherms primarily occur because of the indefinite monolayer-multilayer adsorption process. In monolayer adsorption, the adsorbed molecules are in direct contact with the adsorbent's surface. In contrast, in multilayer adsorption, the adsorption process occurs in more than one layer. So not all the adsorbed molecules are in contact with the adsorbent's surface. This type is commonly found on non-porous or macroporous adsorbents (pore width >50 nm). The sharpness of point B in Fig. 2.1

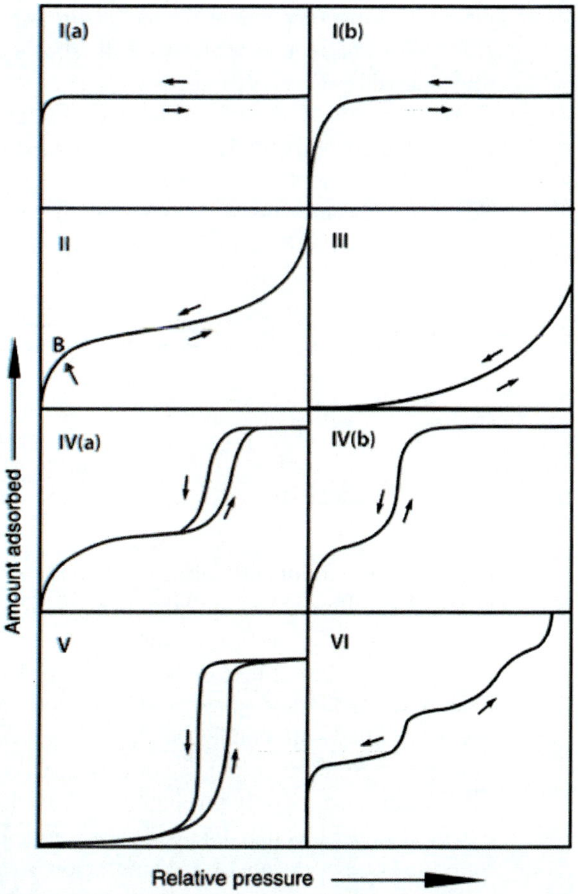


**Fig. 2.1** Types of isotherms classified based on the 2015 IUPAC recommendations [3]

relates to the degree of completion of monolayer adsorption. A more gradual curvature signifies an overlap between monolayer and multilayer adsorption. In Type III isotherms, there is no curvature in the low relative pressure region. Accordingly, there is no monolayer coverage. Although non-porous and macroporous materials show Type III isotherms, this mechanism is characterized by relatively weak adsorbent-adsorbate interactions. Moreover, the adsorbed molecules are clustered around the most favorable sites on the adsorbent's surface.

Type IV isotherms are observed in mesoporous desiccants. The adsorption process takes place in two distinctive stages: monolayer-multilayer adsorption and capillary condensation. Monolayer-multilayer adsorption occurs in the low relative pressure region, and the profile is like that of Type II isotherm. After that, the mechanism is characterized by multilayer adsorption and capillary condensation processes. In capillary condensation, the adsorbate molecules condense to a liquid-like phase in the pores. In the Type IV(a) isotherm, hysteresis is also observed alongside capillary condensation. Such behavior is observed when the pore width surpasses a specific threshold limit (dependent on the adsorbate, adsorbent, and temperature). Adsorbents with smaller mesoporous widths show completely reversible behavior and are represented by Type IV(b) isotherms.

Hydrophobic microporous and mesoporous adsorbents show Type V isotherms. Like Type III, Type V isotherms are characterized by weak adsorbent-adsorbate interaction in the low relative pressure range. The difference between Type V and Type III isotherms emerge at higher relative pressure range and during the desorption process. In Type V, hysteresis is observed during desorption. Lastly, Type VI isotherm is reversible and occurs stepwise. The adsorption process occurs layer-by-layer on a uniform, non-porous surface. The step height denotes each layer's adsorption capacity, and the curvature sharpness signifies the strength of adsorbent-adsorbate interaction.

### *2.2.2 Hysteresis*

Hysteresis predominantly appears in mesoporous materials undergoing multilayer adsorption followed by capillary condensation [5]. Adsorption metastability (a non-equilibrium state that is sometimes perceived as an equilibrium state) or network effect of the pores attribute to the cause of hysteresis. In pores with a cylindrical shape, a delayed condensation process attributes to the adsorption metastability. Since the pores are filled with condensate, thermodynamic equilibrium is established during desorption [6]. Further, in ink-bottle-shaped pores, where wide pores have access to the external surface only through narrow necks, the desorption path depends on the extent of pore blocking. The wide pores remain filled with the adsorbate molecules until the bottlenecks are cleared in the low relative pressure region.

Figure 2.2 shows that the hysteresis loops are classified into six types and are strongly associated with the pore structure of the adsorbents. Types H1 and H4 illustrate the two hysteretic extremes. In Type H1, the adsorption and desorption processes occur vertically and are parallel to each other. Contrastingly, in Type H4, the processes are almost horizontal. Type H2 loops are linked with adsorbents having network effects in the pores. Type H2(a) loop occurs in pores with lower neck widths, whereas Type H2(b) in pores with larger necks. Commonly used mesoporous silica gel desiccant exhibits Type H2 loop while zeolites show Type H4. In Type H3, the adsorption closely resembles Type II isotherm, while hysteresis occurs in desorption due to adsorption metastability, i.e., the presence of partially filled condensates. Compared to other hysteresis types, the occurrence of the Type H5 loop is uncommon since this loop has a distinctive arrangement with both open and partially blocked mesopores.

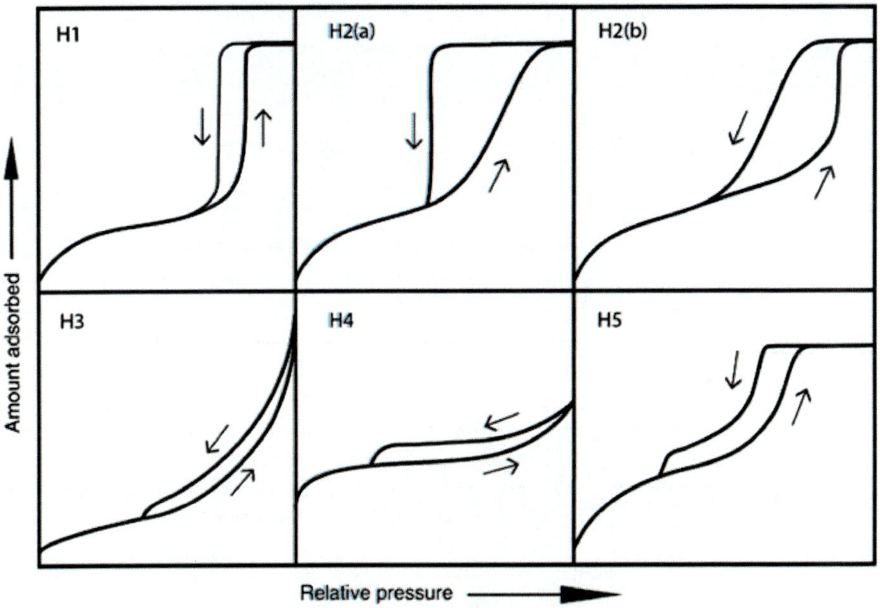


**Fig. 2.2** Types of hysteresis loops classified based on 2015 IUPAC recommendations [3]

## 2.3 Conventional Desiccants

The dehumidification performance of a desiccant dehumidifier depends on the sorption-desorption characteristics of the employed desiccant. This implies that the desiccant must be selected carefully based on its isothermal capacity, kinetics, stability, and regeneration temperature. As noted in the previous section, the desiccant must display a high isothermal capacity since it is directly proportional to the quantity of moisture removed from the air. It should also show fast kinetics as this would determine how quickly the desiccant can capture moisture. Further, the desiccant should remain stable for several thousand cycles, and its regeneration temperature requirement should facilitate effective integration with renewable energy sources.

Figure 2.3 broadly classifies the commonly employed desiccants into three types based on their nature. Ceramic and nanoporous inorganic materials are desiccants with no additional materials that enhance the performance, whereas composite desiccants contain hygroscopic materials to achieve improved dehumidification performance. A detailed review of different types of solid desiccants is presented in the following sections.

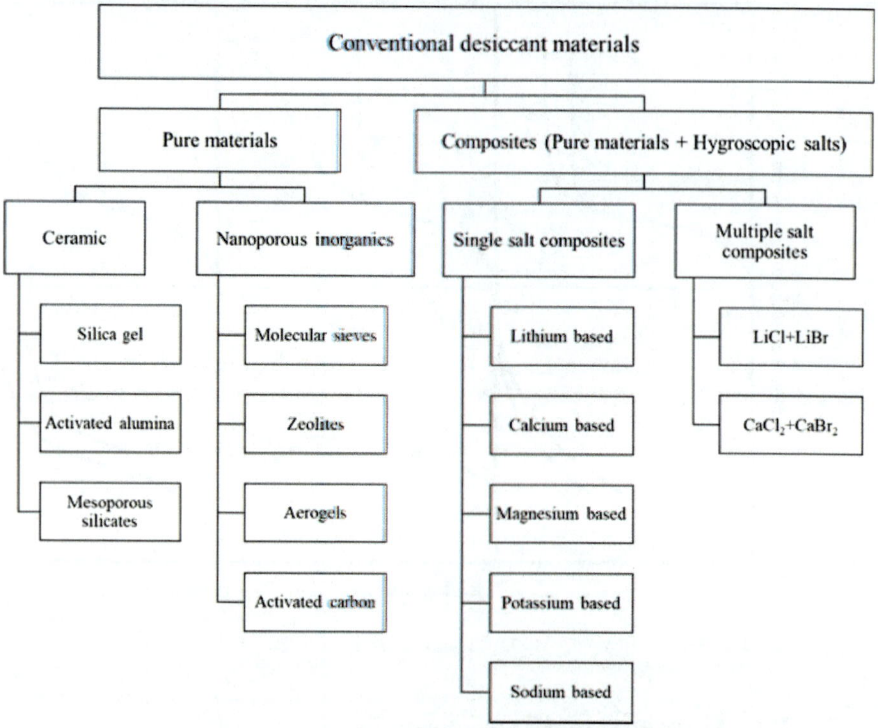


**Fig. 2.3** Classification of the conventional pure/composite desiccant materials based on their natural occurrence

## 2.3.1 *Ceramic Desiccants*

Silica gel, an amorphous form of silicon dioxide ($SiO_2$), is a naturally occurring mineral that is processed and purified into granules, beads, or powders of different particle and pore sizes. On the basis of pore sizes, silica gel can be classified into type A (microporous pores with diameter <2 nm) [7], type B (mesoporous pores with diameter 2–50 nm) [8], and type C (macroporous pores with diameter >50 nm) [9]. Silica gel is one of the most employed materials in desiccant dehumidifiers because it is inexpensive, abundant, hydrophilic, and highly porous with a surface area spanning 300–750 $m^2/g$ [10]. Its maximum isothermal adsorption capacity is approximately 30–40% of its initial dry mass. While its capacity and sizes suit drying and packaging applications, its use in heat transformation applications such as dehumidifiers, heat pumps, and adsorption chillers is severely limited. Due to its low adsorption capacity, higher amount of desiccant is needed for absorbing a fixed quantity of moisture per cycle. Its regeneration temperature requirement is also above 70 °C to achieve appropriate dehumidification capacity [11]. Moreover, silica gel saturates faster, and hence, more frequent regeneration cycles are needed. As its capacity is linked to the surface characteristics, the presence of dust particles and impurities in the pores reduce the moisture adsorption levels. To improve the water adsorption capacity, different concentrations of hygroscopic salts have been combined with silica gel. The significance of synthesizing composite silica gel desiccants will be explored in Sect. 3.3 of this chapter.

Like silica gel, activated alumina is a type of ceramic desiccant. It contains hydrides and oxides of aluminum and is synthesized by dehydrating aluminum hydroxide (purified gibbsite). Its structural characteristics such as pore diameter and surface area can be controlled during its preparation stage by either regulating temperature or heat transfer duration. The pore size range spans 1.5–6 nm and it typically has a higher heat of adsorption, equaling 3000 kJ/kg. In addition, mesoporous silicates are another type of adsorbents synthesized hydrothermally from silica gels. They have highly ordered pore structures, large surface areas (>500 $m^2/g$), and can be chemically modified. Some commonly used mesoporous silicates desiccants are FSM-16, KIT-1, MCM-41, MCM-48, MCM-50, and SBA-15 [12]. The mesoporous silicates have higher water adsorption capacity (up to 80–100% of initial dry mass), however, its regeneration temperature requirement is over 90 °C.

## *2.3.2 Inorganic Desiccants*

Molecular sieves are porous solids with a uniform pore structure and size. It adsorbs molecules smaller than the pore openings and thus acts as a sieve, preventing the adsorption of molecules with larger size. Some molecular sieves have crystalline structures, namely, zeolites, whereas others are amorphous, such as carbon molecular sieves. Since molecular sieves with specific pore size can be synthesized for different heat transformation applications, it is considered as one of the most versatile desiccants [13].

Zeolites are the commonly found molecular sieves, which occur naturally and can also be synthesized chemically. They are complex crystalline aluminosilicates of group IIa and IIIa elements, namely, potassium, sodium, magnesium, and calcium. They are microporous in nature and are based on a three-dimensional interconnecting network of silica and alumina tetrahedrons [14]. There are around 50–70 types of zeolites with different crystal structure and chemical composition. Based on the number of mesh sieves present, zeolites are classified as Type A (4–8 sieves) and Type X (8–12 sieves). Its nomenclature is further based on the material's average pore size: 3A, 4A, 5A, and 13X, etc. [15]. Further, Type Y zeolites (with pore diameter of 0.7–0.8 nm) are hydrophobic materials mainly used in ion-exchange applications [16]. Recently, Mitsubishi plastics developed three highly durable water vapor adsorbing zeolites, namely, AQSOA Z01, Z02, and Z05, based on different isotherm shapes [17, 18]. While they are stable for around 200,000 cycles, it is worthy to note that their adsorption capacity is comparable to that of silica gel.

In addition to the conventional zeolites, zeolite-like molecular sieves such as silicoaluminophosphates (SAPO-34) and ferroaluminophosphates (FAPO-34) are employed in DCHEs [19]. They have high specific surface area (around >500 $m^2/g$)

[20] and record 30–40% higher adsorption capacity than silica gel [19]. While there exist many zeolites with high porosity (defined as the fraction of volume of voids to the total volume of the material) and excellent adsorption potential, the operating conditions required for their synthesis are energy intensive, thereby making them expensive. Another key disadvantage is that they require higher regenerative temperature of the order 120–200 °C. The requirement of heating input at high quality makes the integration with industrial waste heat and solar collectors challenging [21].

Aerogel is another type of nanostructured inorganic porous material with specific surface area of around 600–1000 $m^2/g$ and porosity ranging 0.85–0.99. It is synthesized using silica ($SiO_2$), alumina ($Al_2O_3$), or their mixture, and records around 1.2–1.4 g/g isothermal capacity. Despite the high adsorption capability, its practical utility is largely limited due to a high regeneration temperature requirement of over 80–90 °C [24]. Activated carbon is another type of inorganic desiccant with large pore volume and high surface area (300–4000 $m^2/g$). They are relatively low-cost hydrophobic adsorbents available in powers or beads or fibers forms. It is worth emphasizing that activated carbons are different from carbon molecular sieves based on the surface area and pore size distribution. Activated carbons have a broad range of pores ranging between 1 and 2.5 nm, whereas carbon molecular sieves have narrow pore size distribution, with pore diameters spanning 0.3 and 0.5 nm. In addition, the typical surface area is also significantly smaller (200 and 400 $m^2/g$) for carbon molecular sieves [22].

### *2.3.3 Composite Desiccants: Single Salt and Multi-salt Composites*

The moisture capturing ability of ceramic and inorganic materials is largely limited as the adsorption process is a strong function of the desiccant's pore characteristics. As a result, the conventional pure desiccants demand a larger heat and mass transfer area when employed in dehumidifiers. To overcome this challenge, hygroscopic salts are often impregnated with the existing adsorbents and composite desiccants are accordingly synthesized. Different concentrations of lithium chloride (LiCl) [23], lithium bromide (LiBr), calcium chloride ($CaCl_2$) [24], calcium bromide ($CaBr_2$), calcium nitrate ($Ca(NO_3)_2$) [25], potassium formate ($HCO_2K$) [26], magnesium sulphate ($MgSO_4$) [27], magnesium chloride ($MgCl_2$) [28], and lithium nitrate ($LiNO_3$) [29] have been reported to be embedded. However, the presence of salts causes corrosion and reduces the life cycle of desiccant dehumidifiers. Consequently, the level of salt content is reduced, and only 30–45% improvement in adsorption capacity has been reported. Secondly, upon absorbing water molecules, the salt turns into liquids. This inherent nature washes the desiccant coating off the metallic surfaces or results in the carryover of the salt particles in the supply airstream. The result is that the desiccants' application is adversely impacted.

Like silica gels, mesoporous silicates have also been combined with salts. Two types of SBA-15 desiccant with different pore sizes are combined with $CaCl_2$ and up to 2 times improvements in adsorption capacity can be achieved when compared to the pure silicates. However, the regeneration temperature provision is approximately 90 °C; demonstrating poor energy efficiency. Activated carbon has also been enhanced with hygroscopic salts such as LiCl, $CaCl_2$, sodium silicate ($Na_2SiO_3$) to improve the sorption capacity. Although the isothermal sorption capacity improved by 2–3 times, the composite activated carbon desiccants cannot be effectively coated on heat exchangers due to their low density and poor thermal conductivity which severely limited their performance in desiccant dehumidifiers [30]. Apart from single-salt composites, composite desiccants with multiple salts have been also synthesized, which include LiCl + LiBr and $CaCl_2$ + $CaBr_2$ with silica gel. The binary salt composites demonstrate higher moisture removal as compared to their single salt alternatives.

### 2.3.4 *Limitations of Current Desiccant Materials*

Despite the development of numerous desiccant materials, several drawbacks of conventional pure/composite adsorbents have negatively impacted their performance as dehumidifiers. The most significant challenges associated with existing desiccants are summarized in Table 2.1.

One of the primary limitations of the conventional desiccant is its low isothermal adsorption capacity. As presented in Sect. 2, the isothermal capacity depends on the desiccant's surface and pore characteristics. The constraints in pore characteristics confine the adsorption capacity to 30–40% of the desiccant's dry mass [1]. This drawback continues to be a topic of major interest for several years, and various strategies have been proposed to overcome this drawback. Accordingly, synthesizing composite desiccants by combining pure adsorbents with hygroscopic salts presents an attractive option [31]. While the objective of improving the adsorption capacity can be easily addressed, three critical challenges limit the widespread use of composite desiccants in dehumidifiers:

1. A regeneration temperature higher than 60 °C is required to achieve the desired high dehumidification capacity which affects energy efficiency negatively [19, 32].
2. Higher salt concentrations cause stability-related concerns due to the high probability of washing the desiccant coating away. Accordingly, the system operating time has to be constrained to below 5 min [33], thereby leading to higher desiccant requirements and frequent regenerations.
3. Since the halide salts have demonstrated a strong tendency for corrosion, incorporating them into composite desiccants affects the dehumidifier's durability.

**Table 2.1** Key drawbacks of the conventional pure/composite materials in desiccant dehumidifiers

| Desiccants | Key drawbacks |
|---|---|
| Silica gels | • Low isothermal adsorption capacity<br>• Quick saturation and frequent regeneration<br>• High adsorption heat (2500–2800 $kJkg^{-1}$)<br>• High regeneration temperature (>70 °C) |
| Alumina | • Low isothermal adsorption capacity<br>• Higher regeneration temperature (>70 °C) |
| Mesoporous silicates | • Limited adsorption capacity<br>• Regeneration temperature above 90 °C |
| Zeolites | • High regeneration temperature<br>• High adsorption heat (3000–4500 $kJkg^{-1}$)<br>• Poor thermal conductivity (0.2 $Wm^{-1}K^{-1}$) |
| Activated carbon | • Low thermal conductivity (0.4 $Wm^{-1}K^{-1}$)<br>• High regeneration temperature<br>• Difficult for coating |
| Zeolite-like molecular sieves (SAPO-34 and FAPO-34) | • Limited adsorption capacity<br>• High desiccant synthesis costs<br>• High regeneration temperature |
| Single-salt/Multi-salt composites | • Deliquescence<br>• Carry over of the salt in supply airstream<br>• Corrosion of metallic surfaces<br>• High regeneration temperature |

## 2.4 Binders and Their Selection Criteria

In desiccant dehumidifiers such as rotary wheels or DCHEs, the desiccant material is often impregnated/coated on appropriate substrates such as heat exchangers/foams/plastic sheets. Since ceramic and inorganic pure/composite desiccants have demonstrated poor binding ability, materials with good adhesive properties are necessary to ensure continuous contact of the desiccants with the substrate's surface. Graphite powders [34], polyaniline [35, 36], polyvinyl alcohol (PVA) [37], aluminum hydroxide ($AlOH_3$) [38], emulsion glue [47], and polytetrafluoroethylene (PTFE) [38, 39] have been used to bind silica gel and zeolites on metallic surfaces. In most cases, binder materials with high thermal conductivity are preferred to improve the heat transfer characteristics of the desiccant material. As a result, these binders yielded 3–4 times improvements in the heat transfer rates. Nevertheless, the desiccant's mass transfer characteristics deteriorated significantly. This is because the binder molecules occupied the pores of the adsorbents and impeded the mass transport. Since selecting an appropriate binder is essential in achieving improved dehumidification performance, a comprehensive feasibility study must be carried out. Materials must be analyzed based on these three factors:

(i) Whether the binder demonstrates an excellent adhesive ability.
(ii) If there is any adverse effect on mass transport.
(iii) If the heat transfer rate is enhanced.

The following steps are involved in the binder's selection. A broad list of binder materials that are readily available and cheap must be identified as a first step. Inorganic materials such as bentonite, sepiolite, and silica sols and organic materials such as hydroxy cellulose (HEC), trimethoxypropylsilane, gelatin, and polyvinyl alcohol (PVA) are some of the commonly employed binders. A nitrogen adsorption technique ought to be conducted to measure the binder material's BET surface area [7]. From this test, binders with a high BET area (>300–400 $m^2/g$) can be selected for coating. Next, a homogenous suspension/solution of the desiccant and binder in an appropriate solvent is prepared. A scaled-down prototype of the substrate is then coated with the prepared solution. At this stage, binders demonstrating poor adhesive ability and tending to peel off must be discarded, and materials with excellent binding ability must be chosen for further analysis.

The samples are studied under a universal testing machine (UTM) and field emission scanning electron microscopy (FESEM) to evaluate the binder's coating effectiveness, mechanical strength, and brittleness. A comprehensive list of numerous mechanical tests that can be applied to test the desiccant's binding feasibility is summarized in [40]. To reduce complexity, a more straightforward scrape test can be adopted to check the adhesive ability. A metallic spoon can be employed to scrape the coating layer, and the mass of the desiccant-coated sheet can be measured before and after scraping. If the mass of the desiccant differs marginally, then the binder demonstrates an excellent binding ability. While the scraping method provides a reasonable estimate to check the binding ability, it is worthy to note that it does not replace the need to conduct systematic studies on UTM and FESEM. Figure 2.4 illustrates an example where the coated desiccant mass does not alter significantly (0.26% decrease) after scrape testing.

Once a material's binding capability is established, samples with different binder concentrations are prepared for analyzing the possible detrimental effect on the desiccant's adsorption capacity. BET surface area and pore volumes are measured for the desiccant containing different binder weight ratios. The desiccant-binder configuration displaying the peak value for BET surface area and pore volume can be

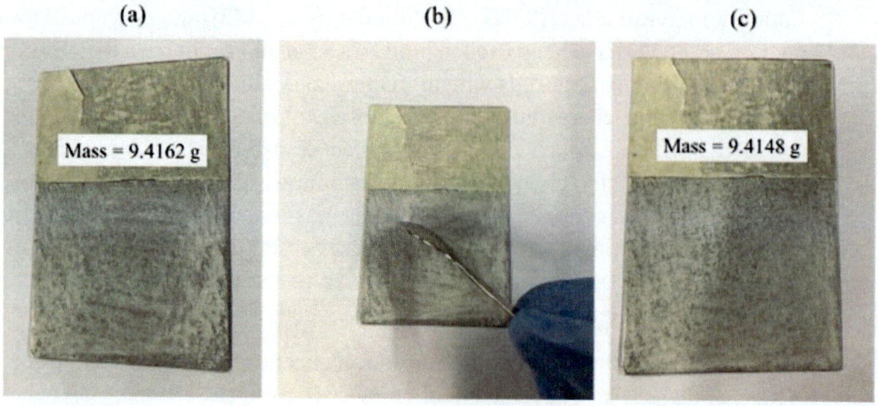


**Fig. 2.4** **a** Desiccant coated aluminum sheet weighing 9.4162 g before scrapping; **b** a metallic spoon is used to scrape the desiccant coating from the surface; and **c** desiccant coated sheet weighing 9.4148 g after scrapping. The initial uncoated sheet weighed 8.8715 g [41]

interpreted as an optimum weight ratio that does not affect the adsorption performance. After obtaining the most effective desiccant-binder concentration, the coated substrate must be exposed to a heat source, and the transient heat transfer coefficient is then evaluated. This step ensures that the heat transfer characteristics of the desiccant are not affected. Figure 2.5 shows the flowchart for selecting appropriate desiccant materials.

So far, a detailed analysis on the selection of binder materials is only available for silica gels. While such a study is necessary for other desiccant materials, it has not been conducted due to the complicated and time-consuming steps involved. With the latest advancements in material science, several new desiccants are synthesized. Therefore, selecting appropriate binders for the advanced desiccants presents a good research gap demanding new fundamental contributions.

## 2.5 Coating Techniques

In stationary bed dehumidifiers, adsorbents are tightly packed on a plate or a heat exchanger (without using any binders). To ensure that the desiccant material remains in contact with its substrate, a wire mesh is used. While this configuration ensures that the desiccant's maximum pore volume is available for moisture removal, due to random particle arrangement and air gaps, the contact area is reduced significantly. Also, the resistance for heat and mass transport is too high, resulting in reduced effectiveness. While the selection process listed in Sect. 4 can be followed to choose an appropriate binder and its concentration, a suitable selected coating procedure is quintessential to achieve a uniform and stable coating. Different techniques,

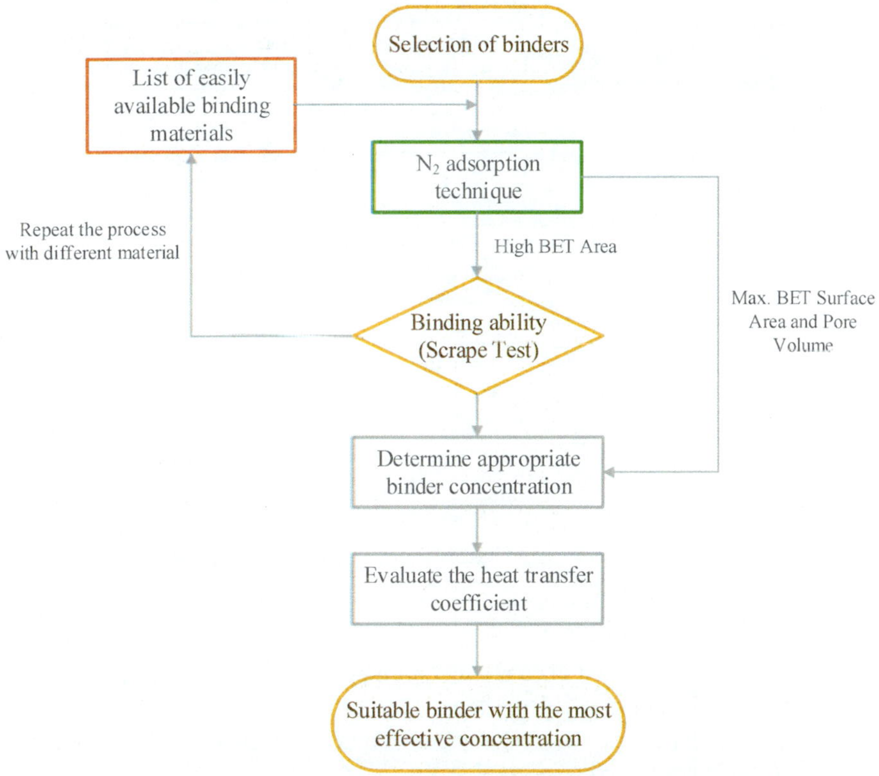


**Fig. 2.5** A flowchart describing the steps involved in selection of a suitable binder material

namely, dip-coating, spray coating, and direct synthesis, have been studied extensively in this regard. The pros and cons of these coating processes are highlighted in the ensuing sub-sections, and Table 2.2 summarizes the comparison between these coating techniques.

### *2.5.1 Dip Coating*

Dip coating is the easiest and one of the least expensive coating techniques. Its key advantage is that it can easily coat substrates of any complex shape while maintaining a uniform coating thickness. In this method, a clean and dry substrate (metal plate/foam/plastic sheet/heat exchanger) on which the desiccant needs to be coated is dipped several times in a homogeneous solution of desiccant and binder. The wet substrate is then placed in an oven at 60–120 °C for 2–4 h or until the substrate is dry. In some cases, to obtain uniform coating layer, the substrate is first placed in a rotating machine so that the dipped desiccant solution is evenly distributed. The rotation machine is controlled using a worm gear and a stepper motor, and its rotation

**Table 2.2** Comparison between the characteristics of different coating processes

| Properties | Dip coating | Spray coating | Direct synthesis |
|---|---|---|---|
| Coating stability | Good | Good | Excellent |
| Coating accuracy | Good | Good | Excellent |
| Reaction conditions | Mild | Mild | Severe |
| Coating thickness (μm) | 120 | 120 | 10–15 |
| Possibility of serial production | Easy | Easy | Difficult |
| Price | Cheap | Moderately Expensive | Expensive |
| Maintenance | Easy | Moderate | Difficult |
| Energy consumption | Low | Moderate | High |

speed is programmed via a microcontroller. The coating process is repeated several times until a uniform coating layer with the desired thickness is obtained [42]. The dip-coating process steps are summarized in a flow chart, as shown in Figs. 2.6, and 2.7 illustrates the process graphically.

It is worthy to note that the solution must have appropriate viscosity to achieve a uniform coating, and the concentration of the solvent must be carefully selected. This is to ensure that the solution is neither too viscous (affecting the flow and blocking the gaps between the fins) nor too dilute (leading to washing away of the desiccant from the substrate). Since the substrate needs to be repeatedly dipped, an uneven dispersion of the desiccant is formed which leads to the blockage of the gaps between the fins. Albeit being simple, the coating method's likely drawbacks include weak mechanical strength, increased volatility, and high heat and mass transfer resistance. Nevertheless, it does not demand any severe reaction conditions and offers direct industrial implementation advantage.

### 2.5.2 *Electrostatic Spray Coating*

Electrostatic spray coater is a dedicated device used to coat powdered desiccants by spraying them onto the surface of the substrate under the influence of an electrostatic field. The powdered desiccant is first loaded into the charging barrel, fluidized in the hopper, and passed to the spray guns. The guns then spray the desiccant powder (up to a coating thickness of ~120 μm) on the surface of the substrate [44]. Figure 2.8 illustrates the spray coating process [45]. This method's key demerits

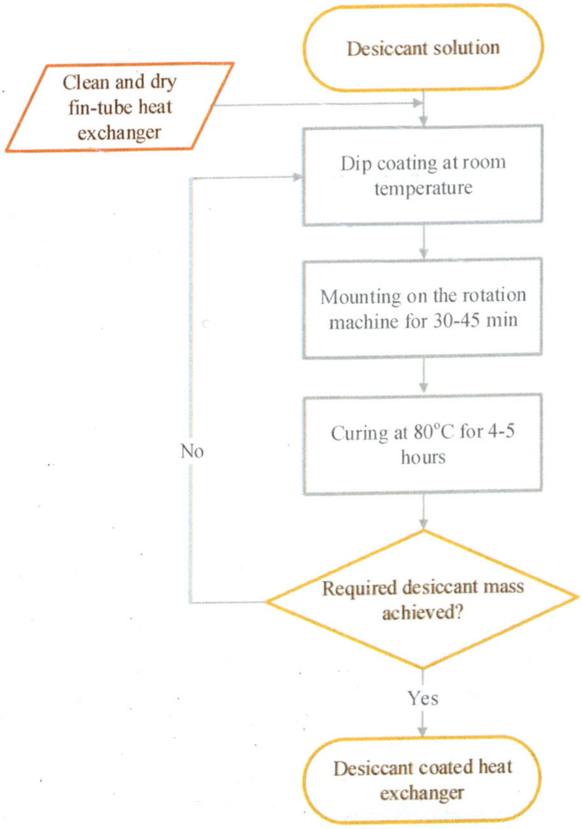


**Fig. 2.6** A flow chart illustrating the steps involved in dip-coating process

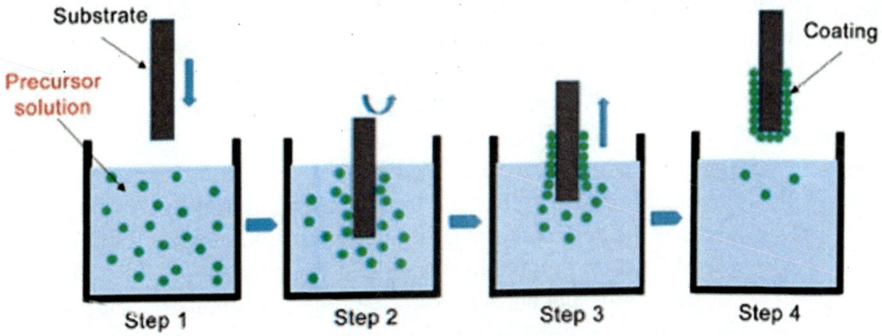


**Fig. 2.7** A schematic representation of the dip-coating process [43]

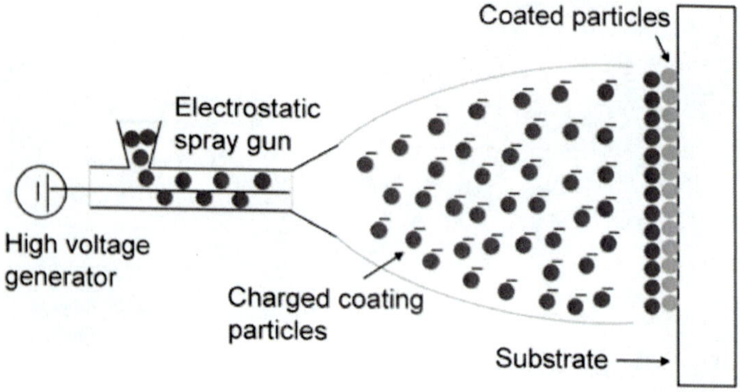


**Fig. 2.8** A schematic illustrating the electrostatic spray coating process [45]

are its high initial investment and frequent maintenance of the movable parts [46]. Despite using a specialized device, no significant benefit is observed in the coating layer obtained through spray coating over the dip-coating process. Also, since the use of powered desiccants is required, it does not offer a straightforward procedure for coating composite desiccants. As composite desiccants are prepared by mixing different materials in a liquid solvent, a higher operation cost is incurred to manufacture dry composite desiccant powders in a large quantity.

### 2.5.3 *Direct Synthesis*

Another technique to carry out coating is by crystallizing the desiccant directly on heat exchanger surface. Different types of zeolites (A, X, and Y) and SAPO-34 have been widely directly synthesized on various metal substrates such as copper, aluminum foams, and stainless steel plates using in situ hydrothermal synthesis process [47, 48]. This process allows zeolite to grow onto the metallic surfaces directly and does not require any binder. The coating is uniform and yields around 10–50 μm thickness. The synthesis takes place in a closed hydrothermal system at a high pressure of around 1 kbar and high-temperature ranges of around 450 °C [49]. Depending on the desiccant used, the synthesis can either last for a few hours or can take several days for completion. Although this process provides excellent coating stability, the synthesis conditions are extreme, and industrialization is practically infeasible [50]. The steps involved in direct synthesis are summarized in Fig. 2.9 [47], and more specific details about the direct-synthesis process can be found in a review article by Capri and co-researchers [51].

Fig. 2.9 A flow chart portraying the steps involved in direct synthesis of desiccants on metallic substrates [47]

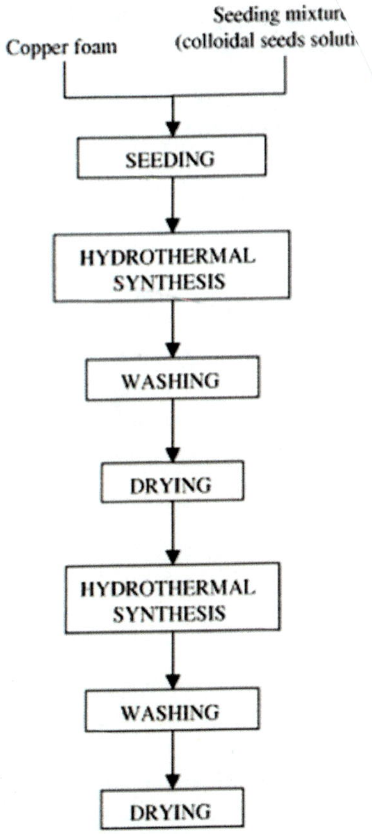


## 2.6 Regeneration Methods

In the previous sections, we comprehensively discussed the different types of desiccants, binders, and coating techniques used in desiccant dehumidifiers. Various strategies employed to achieve improved dehumidification performance have also been highlighted. Irrespective of the amount of moisture that a desiccant is able to capture, the desiccants remove moisture only when the driving force for moisture sorption exists. Once the desiccant has captured sufficient moisture and reached its saturation, reactivating it for subsequent cycles is imperative. This process requires energy input, and it is therefore vital to make regeneration more energy efficient. In Chap. 4, through the second law of thermodynamics approach, the irreversibility associated with the regeneration process will be quantified, and appropriate ways to reduce it will be mentioned. Predominantly, thermal sources such as waste heat/solar thermal collectors/superheated steam have been used; many other techniques have been employed with the motive of achieving improved efficiency [52]. This section reviews the different methods utilized for reactivating desiccants.

### 2.6.1 Thermal Regeneration

As mentioned in Chap. 1, electricity consumption in the air-conditioning application contributes to the most significant proportion of $CO_2e$ emissions. Employing solar energy/thermal waste heat for regenerating a desiccant dehumidifier is attractive since the necessary heating energy is made available without using electricity from the grid. The concept of a solar-powered desiccant cooling system has been researched widely, and many thermal scientists have integrated it with fixed-bed, rotary wheel dehumidifiers, and even with DCHEs. A comprehensive review of solar-assisted air-conditioning technologies has been carried out by Jani and co-researchers [53]. A schematic of the solar thermal collectors powered air-conditioning system employing a desiccant wheel and a DCHE is shown in Fig. 2.10.

Non-concentrated solar thermal technologies such as flat plate collectors (FPCs) and evacuated tube collectors (ETCs) are highly stable and offer higher thermal conversion efficiency [54]. The high regeneration temperature requirement (>80 °C) of many conventional desiccants is currently limiting the application of these solar collectors [55, 56]. Auxiliary electrical water heaters/natural gas boilers are required to raise the water temperature to the desired levels based on the desiccant's regeneration requirement [57, 58]. Such requirements not only reduce energy savings, but also increase system complexity and lower life cycle costs [59, 60]. Since the demand for a higher cooling load is often coincident with the solar irradiation intensity, driving an air-conditioning system via solar energy offers excellent potential to reduce the peak electricity demand and increase the power savings. Consequently, the desiccant cooling system's regeneration temperature requirement must be considerably lowered to facilitate the suitability of non-concentrated solar energy systems.

### 2.6.2 Microwave Regeneration

Microwave regeneration selectively heats the wet desiccant material without increasing the air temperature. Microwaves have been used to regenerate silica gel, activated alumina, activated carbon, zeolite, and molecular sieves [61]. This regeneration method is found to retain the pore structure and original active sites of the desiccants even after many regeneration cycles [62]. While packed-bed systems have been utilized for microwave regeneration, fluidized beds demonstrated better regeneration performance under microwaves. It is because fluidization leads to a more favorable temperature distribution in the desiccant material and facilitates uniform heating. Further, microwave regeneration of a zeolite-based desiccant wheel showed that combining hot air and microwave heating processes improved the regeneration efficiency compared to the standalone hot air regeneration method [63]. A schematic representation of the experimental facility for microwave regeneration in desiccant wheels is shown in Fig. 2.11.

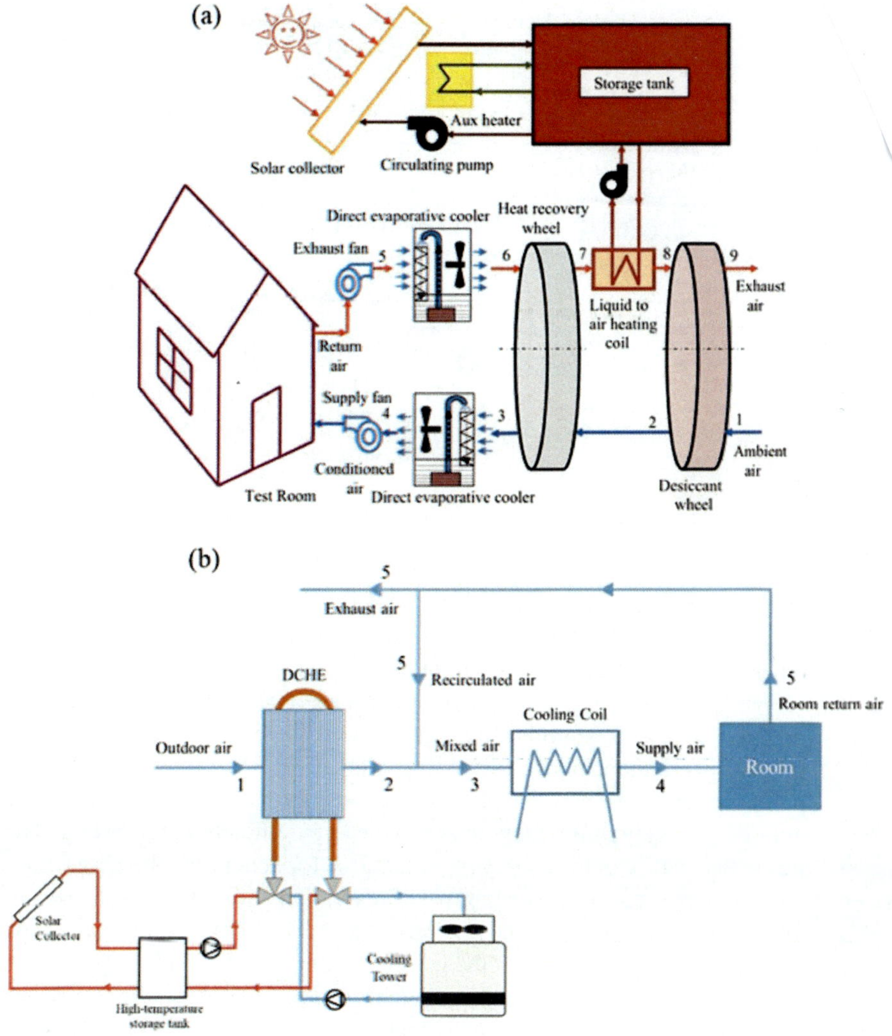


**Fig. 2.10** Solar-assisted air-conditioning system employing **a** desiccant wheels [53] and **b** DCHEs [41]

### *2.6.3 Ultrasound Regeneration*

Ultrasonic waves are characterized by their frequency and intensity. Different acoustic frequencies are combined with power levels to study the regeneration effect of ultrasound. In one of the first studies, a low-frequency, high-intensity ultrasound (20–40 kHz) was transmitted through a silica gel-packed bed [64]. Due to the oscillation of molecules, an alternating compression and expansion effect is generated. The analogy is like squeezing a sponge repeatedly which promotes the desorption of water molecules. Additionally, the desiccant absorbs a part of the ultrasonic energy,

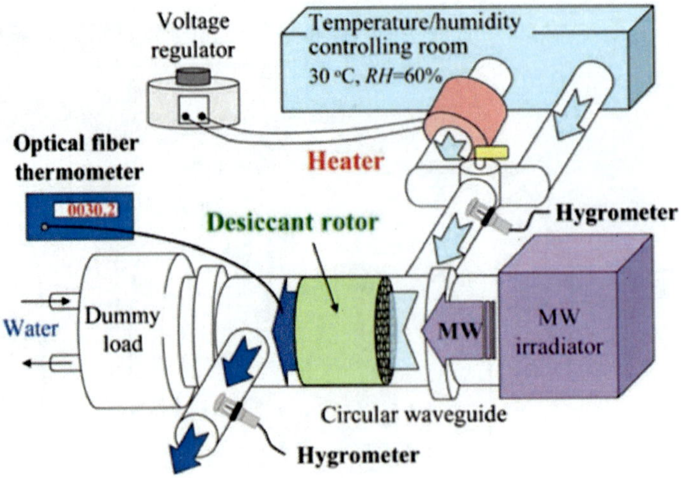


Fig. 2.11 A schematic of microwave regeneration for desiccant wheels [63]

and its temperature rises, resulting in higher mass diffusivity. Thus, the driving force for moisture transfer is enhanced. Figure 2.12 shows the schematic of the different types of experimental facilities employed to conduct the ultrasonic regeneration.

While the method works, there are several unresolved questions related to the underlying physics of ultrasonic regeneration. More studies are needed to establish this regeneration technique as a viable alternative. From an entirely different standpoint, it is also essential to realize why there are so few studies on ultrasonic regeneration even though it claims to achieve energy-efficient regeneration. The answer lies in the quality of the energy input required for generating the ultrasound. In most studies, the power input for thermal regeneration and ultrasonic/microwave regeneration is compared incorrectly. A proper exergetic comparison (via the second law of thermodynamics) between the purely thermal strategy and ultrasonic/microwave regeneration may not yield any significant advantages to these alternative techniques. Lastly, there may exist a severe mismatch due to the acoustic impedance within the air. This may result in almost 99.97% of the ultrasound being reflected at each air interface and minimal energy being available for regeneration. This makes the use of ultrasounds for regenerating desiccant wheels/DCHEs a great topic of rigorous investigation, lest ultrasound is deemed impractical.

## 2.7 Characteristics of an Ideal Desiccant

The desirable properties for an ideal desiccant material for potential use in dehumidification applications are:

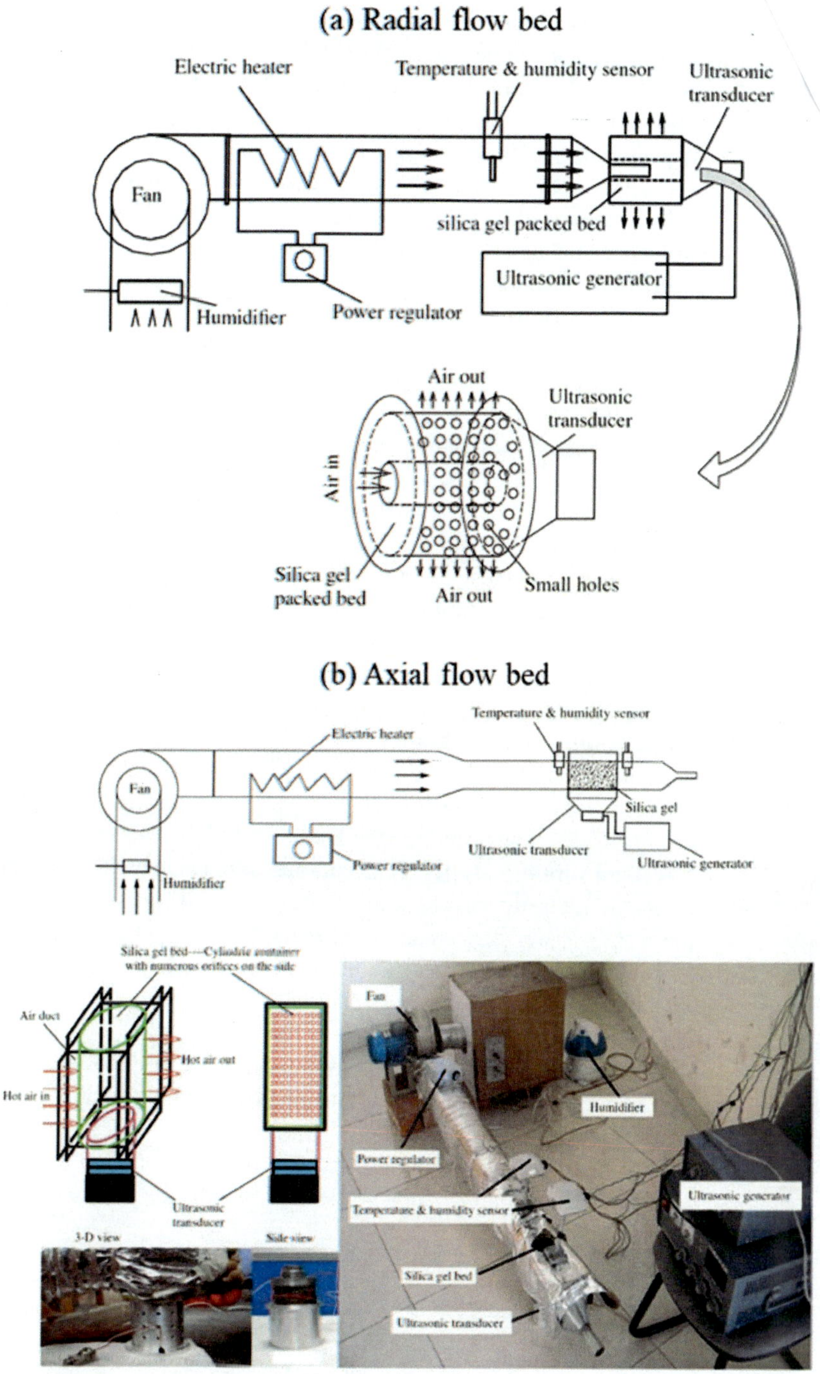


**Fig. 2.12** **a** Radial and **b** Axial flow silica gel packed beds for ultrasonic regeneration

1. **High equilibrium sorption capacity**: A desiccant with high isothermal equilibrium capacity under high RH conditions is required since it reduces the quantity of desiccant needed. As a result, the dehumidifier's footprint and capital expenditure can be lowered.
2. **Appropriate isotherm shape**: Isotherm shape affects the desiccant's suitability for a particular heat transformation application. For desiccant dehumidifiers, a convex-shaped isotherm with lower capacity at low RH and higher capacity under high RH is considered the most effective. However, an S-shaped isotherm is considered appropriate for adsorption chillers.
3. **Heat of sorption**: The desiccant's sorption heat must not change strongly with relative humidity and temperature. It should also be within 5–10% of the heat of vaporization of water. A lower sorption heat ensures that the desiccant's temperature does not increase too high and affects its moisture sorption ability.
4. **Faster kinetics and higher moisture diffusivity**: The transfer of moisture inside the desiccant is a rate-limiting step since the desiccant's moisture diffusivity is 3–4 orders lower than the diffusivity of moisture in the air. As a result, a desiccant with higher moisture diffusivity would allow the desiccant to absorb moisture by its highest potential.
5. **Low regeneration temperature**: A low regeneration temperature (40–60 °C) minimizes the overall energy requirement of desiccant dehumidifiers. In addition, this facilitates easy integration with other cleaner systems such as solar/geothermal energy.
6. **Good stability and durability**: The desiccant should be capable of undergoing thousands of dehumidification-regeneration cycles without losing its moisture capturing ability. This factor lowers the overall maintenance requirements and minimizes any downtime-related losses.
7. **Excellent binding ability**: Conventional desiccants have poor binding ability on metallic/plastic surface which forces the need for adhesives. This reduces the mass transfer performance and may affect the dehumidifier performance. If the desiccant has intrinsic ability to act as an adhesive, then there would be no need for additional binding materials.
8. **No deliquescence and carry over**: The possibility of deliquescence causes the desiccant coating to be wash away. Therefore, any possibility of liquification of the desiccant must be controlled by regulating the hygroscopic salt's concentration.
9. **Non-corrosive:** Corrosion affects the dehumidification performance and requires higher maintenance costs. As a result, the desiccant should not be prone to corrosion.
10. Lastly, the desiccant must be cheap, non-toxic, and non-reactive.

## 2.8 Conclusions

This chapter presents the current state-of-the-art desiccants, binders, coating techniques, and regeneration methods. Firstly, different types of water adsorption isotherms are described, and commonly found pure/composite ceramic and inorganic desiccants are reviewed. The composite ceramic/inorganic desiccants yielded better isothermal capacity than the pure desiccants. However, key challenges due to deliquescence, durability, and higher regeneration temperature requirements made their practical application challenging. Secondly, different types of binding materials utilized to achieve good contact with a substrate are highlighted. Most of the binders negatively affected the mass transfer performance since they block the adsorbent materials' pores. Consequently, a judicious selection of binders with excellent adhesive properties and improved heat and mass transfer rates are desired. Further, different ways for coating and regenerating the desiccants are also described. Among the three coating techniques, the dip-coating method offered simple design, cost-effectiveness, and serial production ability. It was found that adequate research studies on microwave and ultrasonic regeneration techniques are not available; therefore, thermal regeneration is still an attractive option. However, the regeneration temperature of conventional desiccants must be reduced to 40–50 °C from the current requirement of greater than 80 °C to facilitate the effective integration of solar-powered devices. These factors pave the way for synthesizing advanced materials that can achieve superior energy efficiencies.

## References


1. Vivekh P, Kumja M, Bui DT, Chua KJ. Recent developments in solid desiccant coated heat exchangers—a review. Appl Energy. 2018;229:778–803. https://doi.org/10.1016/j.apenergy.2018.08.041.
2. Sing KS, Everett DH, Haul RAW, Moscou L, Pierotti RA, Rouquerol J, et al. Reporting physisorption data for gas/solid systems-with special reference to the determination of surface area and porosity. Pure Appl Chem. 1985;57:603–19.
3. Thommes M, Kaneko K, Neimark A V, Olivier JP, Rodriguez-Reinoso F, Rouquerol J, et al. IUPAC Technical Report Physisorption of gases, with special reference to the evaluation of surface area and pore size distribution (IUPAC Technical Report). Pure Appl Chem 2015. https://doi.org/10.1515/pac-2014-1117.
4. Dubinin MM. Adsorption in micropores. J Colloid Interface Sci. 1967;23:487–99. https://doi.org/10.1016/0021-9797(67)90195-6.
5. Donohue MD, Aranovich GL. Adsorption hysteresis in porous solids. J Colloid Interface Sci. 1998;205:121–30. https://doi.org/10.1006/jcis.1998.5639.
6. Pan G, Liss PS. Metastable-equilibrium adsorption theory: I theoretical. J Colloid Interface Sci. 1998;201:71–6. https://doi.org/10.1006/JCIS.1998.5396.
7. Li A, Thu K, Ismail A Bin, Shahzad MW, Ng KC. Performance of adsorbent-embedded heat exchangers using binder-coating method. Int J Heat Mass Transf 2016;92:149–57. https://doi.org/10.1016/j.ijheatmasstransfer.2015.08.097.

8. Ismail A Bin, Li A, Thu K, Ng KC, Chun W. On the Thermodynamics of Refrigerant + Heterogeneous Solid Surfaces Adsorption. Langmuir 2013;29:14494–502. https://doi.org/10.1021/la403330t.
9. Ge TS, Dai YJ, Wang RZ, Peng ZZ. Experimental comparison and analysis on silica gel and polymer coated fin-tube heat exchangers. Energy. 2010;35:2893–900. https://doi.org/10.1016/j.energy.2010.03.020.
10. Christy AA. Quantitative determination of surface area of silica gel particles by near infrared spectroscopy and chemometrics. Colloids Surfaces A Physicochem Eng Asp. 2008;322:248–52. https://doi.org/10.1016/j.colsurfa.2008.03.021.
11. Vivekh P, Bui DT, Wong Y, Kumja M, Chua KJ. Performance evaluation of PVA-LiCl coated heat exchangers for next-generation of energy-efficient dehumidification. Appl Energy 2019;237. https://doi.org/10.1016/j.apenergy.2019.01.018.
12. Alothman ZA. a review: fundamental aspects of silicate mesoporous materials 2012:2874–902. https://doi.org/10.3390/ma5122874.
13. Flanigen EM. Chapter 2 Zeolites and Molecular Sieves an Historical Perspective. Stud Surf Sci Catal 1991;58:13–34. https://doi.org/10.1016/S0167-2991(08)63599-5.
14. Maesen T, Marcus B. Chapter 1 The zeolite scene—An overview. Stud Surf Sci Catal 2001;137:1–9. https://doi.org/10.1016/S0167-2991(01)80242-1.
15. Type A, X Zeolites n.d. https://www.acsmaterial.com/blog-detail/type-a-and-x-zeolites.html. Accessed March 11, 2021.
16. Y-type Zeolites n.d. https://www.acsmaterial.com/blog-detail/y-type-zeolites.html. Accessed March 11, 2021.
17. Mitsubishi Plastics. Zeolitic Water Vapor Adsorbent AQSOA™. 2008.
18. Mitsubishi Plastics to Test-Market Easy-to-Install Compact Adsorption Chiller with Integrated Cooling Tower to the U.S. Market | Business Wire n.d. https://www.businesswire.com/news/home/20120117006242/en/Mitsubishi-Plastics-Test-Market-Easy-to-Install-Compact-Adsorption-Chiller. Accessed April 4, 2018.
19. Zheng X, Wang RZ, Ge TS, Hu LM. Performance study of SAPO-34 and FAPO-34 desiccants for desiccant coated heat exchanger systems. Energy. 2015;93:88–94. https://doi.org/10.1016/j.energy.2015.09.024.
20. SAPO-34 | Molecular Sieves | ACS Material, LLC n.d. https://www.acsmaterial.com/sapo-34.html. Accessed March 11, 2021.
21. Munz GM, Bongs C, Morgenstern A, Lehmann S, Kummer H, Henning HM, et al. First results of a coated heat exchanger for the use in dehumidification and cooling processes. Appl Therm Eng. 2013;61:878–83. https://doi.org/10.1016/j.applthermaleng.2013.08.041.
22. Lemcoff NO. Nitrogen separation from air by pressure swing adsorption. vol. 120. Elsevier Masson SAS; 1998. https://doi.org/10.1016/S0167-2991(99)80557-6.
23. Jiang Y, Ge TS, Wang RZ, Hu LM. Experimental investigation and analysis of composite silica-gel coated fin-tube heat exchangers. Int J Refrig. 2015;51:169–79. https://doi.org/10.1016/j.ijrefrig.2014.11.012.
24. Leiming H, Tianshu G, Yu J, Ruzhu W. Hygroscopic property of metal matrix composite desiccant. J Refrig. 2014;2:12.
25. Simonova IA, Freni A, Restuccia G, Aristov YI. Water sorption on composite "silica modified by calcium nitrate". Microporous Mesoporous Mater. 2009;122:223–8. https://doi.org/10.1016/j.micromeso.2009.02.034.
26. Ge TS, Zhang JY, Dai YJ, Wang RZ. Experimental study on performance of silica gel and potassium formate composite desiccant coated heat exchanger. Energy. 2017;141:149–58. https://doi.org/10.1016/j.energy.2017.09.090.
27. Gordeeva LG, Glaznev I, Aristov YI. Sorption of water by sodium, copper, and magnesium sulfates dispersed into mesopores of silica gel and alumina. Undefined 2003.
28. Zheng X, Ge TS, Wang RZ. Recent progress on desiccant materials for solid desiccant cooling systems. Energy. 2014;74:280–94. https://doi.org/10.1016/j.energy.2014.07.027.
29. Sapienza A, Glaznev IS, Santamaria S, Freni A, Aristov YI. Adsorption chilling driven by low temperature heat: new adsorbent and cycle optimization. Appl Therm Eng. 2012;32:141–6. https://doi.org/10.1016/j.applthermaleng.2011.09.014.

30. Zheng X, Wang RZ, Ge TS. Experimental study and performance predication of carbon composite desiccants for desiccant coated heat exchangers. Int J Refrig. 2016;72:124 https://doi.org/10.1016/j.ijrefrig.2016.03.013.
31. Hu LM, Ge TS, Jiang Y, Wang RZ. Performance study on composite desiccant material coate fin-tube heat exchangers. Int J Heat Mass Transf 2015;90:109–20. https://doi.org/10.1016/j.ijheatmasstransfer.2015.06.033.
32. He ZH, Huang HY, Lin ZX, Yuan HR, Kobayashi N, Guo HF, et al. Study on adsorption desiccant based hybrid air conditioning system. Electr Power Energy Syst vol 516, Trans Tech Publications; 2012, pp. 1196–200. https://doi.org/10.4028/www.scientific.net/AMR.516-517.1196.
33. Ge TS, Dai YJ, Wang RZ. Analysis on integrated low grade condensation heat powered desiccant coated vapor compression system. Appl Therm Eng. 2018;138:307–18. https://doi.org/10.1016/j.applthermaleng.2018.04.044.
34. Yanagi H, Ino N. Heat and mass transfer characteristics in consolidated silica gel/water adsorption–cooling system. ASME 1997 Turbo Asia Conf., American Society of Mechanical Engineers; 1997, p. V001T13A009-V001T13A009.
35. Hu EJ, Zhu D-S, Sang X-Y, Wang L, Tan Y-K. Enhancement of thermal conductivity by using polymer-zeolite in solid adsorption heat pumps. J Heat Transfer. 1997;119:627–9. https://doi.org/10.1115/1.2824151.
36. Wang L, Zhu D, Tan Y. Heat transfer enhancement on the adsorber of adsorption heat pump. Adsorption. 1999;5:279–86.
37. Chang K-S, Chen M-T, Chung T-W. Effects of the thickness and particle size of silica gel on the heat and mass transfer performance of a silica gel-coated bed for air-conditioning adsorption systems. Appl Therm Eng. 2005;25:2330–40.
38. Pino L, Aristov YI, Cacciola G, Restuccia G. Composite materials based on zeolite 4A for adsorption heat pumps. Adsorption. 1997;3:33–40. https://doi.org/10.1007/BF01133005.
39. Basile A, Cacciola G, Colella C, Mercadante L, Pansini M. Thermal conductivity of natural zeolite-PTFE composites. Heat Recover Syst CHP. 1992;12:497–503. https://doi.org/10.1016/0890-4332(92)90018-D.
40. Freni A, Frazzica A, Dawoud B, Chmielewski S, Calabrese L, Bonaccorsi L. Adsorbent coatings for heat pumping applications: Verification of hydrothermal and mechanical stabilities. Appl Therm Eng. 2013;50:1658–63. https://doi.org/10.1016/j.applthermaleng.2011.07.010.
41. Vivekh P, Islam MR, Chua KJ. Experimental performance evaluation of a composite superabsorbent polymer coated heat exchanger based air dehumidification system. Appl Energy. 2020;260: https://doi.org/10.1016/j.apenergy.2019.114256.
42. Oh SJ, Ng KC, Chun W, Chua KJE. Evaluation of a dehumidifier with adsorbent coated heat exchangers for tropical climate operations. Energy. 2017. https://doi.org/10.1016/j.energy.2017.02.169.
43. Neacşu IA, Nicoară AI, Vasile OR, Vasile BŞ. Inorganic micro- and nanostructured implants for tissue engineering. Nanobiomaterials Hard Tissue Eng Appl Nanobiomaterials, Elsevier Inc.; 2016, p. 271–95. https://doi.org/10.1016/B978-0-323-42862-0.00009-2.
44. Fang Y, Zuo S, Liang X, Cao Y, Gao X, Zhang Z. Preparation and performance of desiccant coating with modified ion exchange resin on finned tube heat exchanger. Appl Therm Eng. 2016;93:36–42. https://doi.org/10.1016/j.applthermaleng.2015.09.044.
45. Nagarajan B, Hu Z, Song X, Zhai W, Wei J. Development of micro selective laser melting: the state of the art and future perspectives. Engineering. 2019;5:702–20. https://doi.org/10.1016/j.eng.2019.07.002.
46. Powder Coatings Electrostatic vs Fluid Bed n.d. http://www.coatings.org.uk/Sectors/Powder_Coatings_Electrostatic_vs_Fluid_Bed_.aspx. Accessed March 13, 2017.
47. Bonaccorsi L, Freni A, Proverbio E, Restuccia G, Russo F. Zeolite coated copper foams for heat pumping applications. Microporous Mesoporous Mater. 2006;91:7–14. https://doi.org/10.1016/j.micromeso.2005.10.045.

48. Schnabel L, Tatlier M, Schmidt F, Erdem-Şenatalar A. Adsorption kinetics of zeolite coatings directly crystallized on metal supports for heat pump applications (adsorption kinetics of zeolite coatings). Appl Therm Eng. 2010;30:1409–16. https://doi.org/10.1016/j.applthermaleng.2010.02.030.
49. Petrov I, Michalev T. Synthesis of Zeolite A: A Review. НАУЧНИ ТРУДОВЕ НА РУСЕНСКИЯ УНИВЕРСИТЕТ (Proceedings - Chem Technol 2012:30–5.
50. Freni A, Bonaccorsi L, Calabrese L, Caprì A, Frazzica A, Sapienza A. SAPO-34 coated adsorbent heat exchanger for adsorption chillers. Appl Therm Eng. 2015;82:1–7. https://doi.org/10.1016/j.applthermaleng.2015.02.052.
51. Caprì A, Frazzica A, Calabrese L. Recent developments in coating technologies for adsorption heat pumps: A review. Coatings. 2020;10:1–24. https://doi.org/10.3390/coatings10090855.
52. Shamim JA, Hsu WL, Paul S, Yu L, Daiguji H. A review of solid desiccant dehumidifiers: current status and near-term development goals in the context of net zero energy buildings. Renew Sustain Energy Rev. 2021;137: https://doi.org/10.1016/j.rser.2020.110456.
53. Jani DB, Mishra M, Sahoo PK. Solid desiccant air conditioning—a state of the art review. Renew Sustain Energy Rev. 2016;60:1451–69. https://doi.org/10.1016/j.rser.2016.03.031.
54. Kalogirou SA. Solar energy engineering: processes and systems. 3rd edn. Elsevier Inc.; 2014. https://doi.org/10.1016/C2011-0-07038-2.
55. Eicker U, Schneider D, Schumacher J, Ge T, Dai Y. Operational experiences with solar air collector driven desiccant cooling systems. Appl Energy. 2010. https://doi.org/10.1016/j.apenergy.2010.06.022.
56. Beccali M, Finocchiaro P, Nocke B. Energy and economic assessment of desiccant cooling systems coupled with single glazed air and hybrid PV/thermal solar collectors for applications in hot and humid climate. Sol Energy. 2009;83:1828–46. https://doi.org/10.1016/j.solener.2009.06.015.
57. Heidari A, Roshandel R, Vakiloroaya V. An innovative solar assisted desiccant-based evaporative cooling system for co-production of water and cooling in hot and humid climates. Energy Convers Manag. 2019;185:396–409. https://doi.org/10.1016/j.enconman.2019.02.015.
58. White SD, Lin S, Guo J, Sproul AB, Bilbao JI. A review of photovoltaic thermal (PV/T) heat utilisation with low temperature desiccant cooling and dehumidification. Renew Sustain Energy Rev. 2016;67:1–14. https://doi.org/10.1016/j.rser.2016.08.056.
59. Jani DB, Mishra M, Sahoo PK. A critical review on application of solar energy as renewable regeneration heat source in solid desiccant—vapor compression hybrid cooling system. J Build Eng. 2018;18:107–24. https://doi.org/10.1016/j.jobe.2018.03.012.
60. Jani DB, Mishra M, Sahoo PK. Performance studies of hybrid solid desiccant-vapor compression air-conditioning system for hot and humid climates. Energy Build. 2015;102:284–92. https://doi.org/10.1016/j.enbuild.2015.05.055.
61. US4805317A—Microwave regeneration of adsorbent materials for reuse as drying agents—Google Patents n.d. https://patents.google.com/patent/US4805317A/en. Accessed March 15, 2021.
62. Foo KY, Hameed BH. Microwave-assisted regeneration of activated carbon. Bioresour Technol. 2012;119:41–7. https://doi.org/10.1016/j.biortech.2012.05.061.
63. Kubota M, Hanada T, Yabe S, Matsuda H. Regeneration characteristics of desiccant rotor with microwave and hot-air heating. Appl Therm Eng. 2013;50:1576–81. https://doi.org/10.1016/j.applthermaleng.2011.11.044.
64. Yao Y, Zhang W, Peng Y, Wang L, Liu Y, Chen B. Modeling of silica gel dehydration assisted by power ultrasonic * corresponding author : Ye Yao. Refrig Air Cond 2010;1–9.

# Chapter 3
# Latest Developments in the Desiccant-Coated Dehumidifiers

Check for updates

**Abstract** The performance of desiccant-coated heat exchangers (DCHEs) strongly depends on the sorption and desorption characteristics of the employed desiccant material. While several attempts have been made to synthesize superior desiccants, the DCHE performance has not reached its highest potential due to challenges associated with the limited working capacity of the conventional pure/composite desiccants. This chapter presents two novel advanced desiccants, namely, composite superabsorbent polymers (SAPs) and metal-organic frameworks (MOFs). The composite SAPs offer superior sorption capacity, faster kinetics, and low regeneration possibility while the MOFs providing excellent hydrophilicity and tailorable structures. These desiccants are deemed to be the next generation of advanced materials in thermally driven dehumidifiers. Aside from documenting the characteristics of these new materials, a detailed list of experimental and theoretical material characterization studies is provided. Isotherms and kinetics are identified as key properties that fundamentally govern the desiccant's performance in dehumidifiers, and relevant experimental and regression techniques to analyze them are also presented. In addition, the transient performance of the superabsorbent polymer DCHEs is compared and benchmarked against silica gel-coated heat exchangers. Lastly, results from a series of parametric experiments are presented by varying different operating parameters, and their sensitivity towards dehumidification capacity and thermal energy efficiency is estimated.

**Keywords** Superabsorbent polymers · Metal-organic frameworks · Characterization · Kinetics · Experiments · Performance analysis

## List of Symbols

| | |
|---|---|
| $A$ | Pre-exponential factor in Arrhenius equation, 1/s |
| $c$ | Constant in Clausius Clapeyron equation |
| $c_p$ | Specific heat capacity at constant pressure, $kJkg^{-1}K^{-1}$ |
| $d_{pipe}$ | Diameter of pipe where air velocity meter is mounted |
| $E_a$ | Activation energy, kJ/mol |

(continued)


V. Prabakaran and K. J. Chua, *Advances in Desiccant Dehumidification*,
https://doi.org/10.1007/978-3-030-80843-3_3

| | |
|---|---|
| $f$ | Mathematical relationship between the derived and measured parameters |
| $H$ | Height, m |
| $\Delta H_s$ | Heat of sorption, kJ/kg |
| $\Delta H_{vap}$ | Heat of vaporization, kJ/kg |
| $k$ | Kinetic constant, 1/s |
| $L$ | Length, m |
| $m$ | Mass, g or kg |
| $\dot{m}$ | Mass flow rate, kg/s |
| $n$ | Number of heat exchangers/number of parameters directly measured Using the sensors |
| $P$ | Pressure, kPa |
| $q$ | Water uptake capacity, kg/kg |
| $\dot{Q}$ | Cooling load, kJ/kg |
| $R$ | Ideal gas constant, $\text{kJmol}^{-1}\text{K}^{-1}$ |
| $R^2$ | Coefficient of determination |
| $T$ | Absolute temperature, °C |
| $t_{a,int}$ | Interaction time between air and desiccant, s |
| $t_{cyc}$ | Cycle time |
| $V_f$ | Face velocity of air at the heat exchanger inlet, m/s |
| $V_{sen}$ | Velocity of air measured by the air velocity meter |
| $W$ | Width, m |
| $x$ | Measured parameters |
| $y$ | Derived parameters |
| **Greek symbols** | |
| $\delta$ | Uncertainty |
| $\Delta$ | Change |
| $\rho$ | Density, kg/m$^3$ |
| $\omega$ | Humidity ratio, g/kg |
| **Subscripts** | |
| 0 | Initial state |
| a | Air |
| atm | Atmosphere |
| cw | Cooling water |
| d | Dry desiccant |
| db | Dry bulb |
| e | Equilibrium |
| hw | Hot water |
| in | Inlet |

(continued)

| | |
|---|---|
| l | Latent |
| out | Outlet |
| s | Sensible |
| sat | Saturated |
| t | Time/total |
| v | Vapor |
| w | Water/water vapor/wet desiccant |
| wb | Wet bulb |

**Abbreviations**

| | |
|---|---|
| $COP_{th}$ | thermal coefficient of performance |
| LDF | linear driving force |
| MRC | moisture removal capacity, g/kg |
| NRMSE | normalized root mean squared error |
| PLCHE | PVA-LiCl (50 w%) coated heat exchanger |
| RH | relative humidity, % |
| SGCHE | silica gel coated heat exchanger |
| SLCHE | SAP-LiCl (50 w%) coated heat exchanger |
| SPFCHE | SAP-$HCO_2K$ (50 w%) coated heat exchanger |

## 3.1 Introduction

A review of many pure/composite ceramic and inorganic desiccants was conducted in the previous chapter. Their applicability in desiccant dehumidifiers were challenged by low sorption capacity, slower kinetics, deliquescence, frequent switching, and high regeneration temperature that have not been addressed completely. Recent developments in material science have expanded the scope for synthesizing next-generation desiccants and addressed these research gaps. This chapter highlights such advances and presents the developments in composite polymers and metal-organic-frameworks (MOFs). A detailed experimental methodology is then introduced to evaluate the sorption-desorption characteristics. Lastly, the desiccant's dynamic performance is judiciously analyzed when coated on a fin-tube heat exchanger.

## 3.2 Next-Generation Advanced Desiccants

### *3.2.1 Superabsorbent Polymer-Based Composites*

In the late 1960s, the agriculture department of the United States government developed new cross-linked polymers using acrylates and acrylamides. They could hold water up to 300–400 times their weight and were readily employed in agriculture and diaper industries [1]. These superabsorbent polymers (SAP) were thought to pave the way as the next-generation desiccant materials. Accordingly, around 30 unique SAPs were synthesized with amines and acrylates [2, 3]. Although these cross-linked polymers recorded excellent isothermal sorption capacity, they degraded rapidly, which implied these polymers did not even achieve short-to-medium term stability. Due to this reason, not many research initiatives were carried out to study such polymers as desiccants and even the causes of their instability remained unexplored [4, 5].

Recently, several scientists have reinitiated their research efforts to study SAPs as desiccant materials. White and co-researchers [6] were among the first groups to perform a comparative study on the dehumidification performance of silica gel, zeolite, and SAPs. Their experimental analysis demonstrated that the SAP-based desiccant wheel performed much better than silica gel and zeolites. Further, they showed that SAPs could be regenerated at low regeneration temperatures of less than 80 °C when 80–120 °C were necessary for the conventional desiccants. In another study, Higashi and co-workers [7] coated SAP on fin-tube heat exchangers and exhibited that the moisture diffusivity in SAP improved by 1–2 orders of magnitude compared to silica gel coated heat exchangers. The stability aspect of the polymer desiccants was also analyzed via an experimental study conducted by Lee and the team [8]. They impregnated hygroscopic salts with SAP and developed a composite polymer desiccant suitable for desiccant wheels. Improvement in sorption capacity of 2–3 times could be achieved. Further, the composite SAP desiccant wheel was able to operate for over 40,000 cycles without any deterioration in its moisture capturing ability. Yang and co-workers [9] went a step further and showed that the presence of LiCl in SAP contributed to better dehumidification performance, and 2 times higher sorption capacity could be achieved with 10.6 w% LiCl solution in SAP than pure SAP desiccant. In addition, several research initiatives employed composite PVA desiccant with LiCl in membrane dehumidifiers due to its high water sorption capability [10–13]. The observations from these studies converge toward an excellent potential for employing composite superabsorbent polymer materials in desiccant dehumidifiers. The development of composite polymer desiccants along with its experimental performance evaluation in desiccant dehumidifier will be discussed in detail in Sects. 3.3 and 3.4 of this chapter.

### 3.2.2 *Metal-Organic Frameworks (MOFs)*

Besides SAPs, MOFs have emerged strongly as superior materials for desiccant dehumidification applications [14, 15]. These porous materials show advanced chemical and structural tunability. Constructed from organic struts and metal ions/clusters, their precise hydrophilicity control allows appropriate interaction with the water molecules [16]. Yet, the metal ions/clusters are vulnerable to hydrolysis in the presence of water, as depicted in Fig. 3.1 [17]. However, if hydrophobic linkers or stable secondary building units (SBUs) are employed, the hydrolytic stability of MOFs can be markedly improved [18].

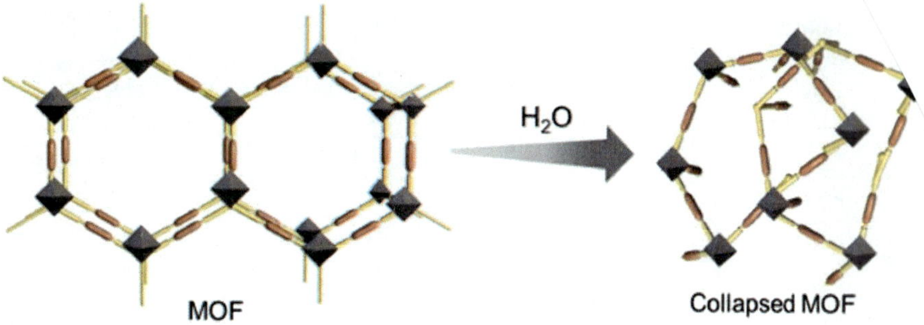


**Fig. 3.1** A schematic showing ordered MOF collapsing in the presence of water [17]

There are over several thousand of MOF structures available in the literature, which makes their selection extremely challenging [17, 19]. Thus far, only a handful of MOFs have been employed in moisture sorption applications, and illustrations of a few water-stable MOFs are shown in Fig. 3.2. In addition, a list of selected MOFs with a considerable water adsorption capacity is summarized in Table 3.1. Besides the ideal characteristics of a desiccant (Sect. 2.7), other key points that need to be considered while selecting MOFs are, (a) the MOF must be stable upon hydrolysis,

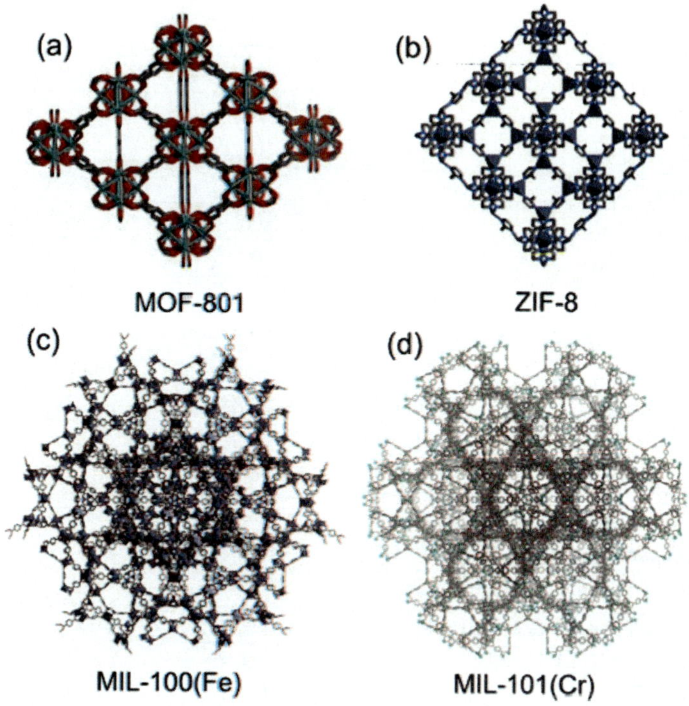


**Fig. 3.2** Some water-stable MOFs reported in the literature [17]

**Table 3.1** A summary of the water uptake capacity and pore characteristic properties of some well-known MOFs

| MOF | BET surface area ($m^2/g$) | Pore volume ($cm^3/g$) | Water uptake capacity (g/g) | Literature |
|---|---|---|---|---|
| CAU-10 | 635 | 0.25 | 0.31 | [14] |
| CAU-10-H | 635 | 0.5 | 0.382 | [21] |
| CAU-10-$NH_2$ | N.A. | N.A. | 0.19 | [14] |
| CAU-10-$OCH_3$ | N.A. | N.A. | 0.15 | [14] |
| CAU-6 | 620 | 0.25 | 0.485 | [14] |
| DUT-67 | 1560 | 0.60 | 0.625 | [22] |
| MIL-100(Fe) | 1549 | 0.82 | 081 | [23] |
| MIL-100(Cr) | 1517 | 0.77 | 0.40 | [23] |
| MIL-100(Al) | 1814 | 1.14 | 0.50 | [23] |
| MIL-101(Cr) | 3000–4000 | 1.8 | 1.4 | [24] |
| MIL-101(Cr)-$NH_2$ | 2690 | 1.27 | 1.06 | [25] |
| MIL-101(Cr)-$NO_2$ | 2146 | 1.19 | 1.08 | [25] |
| MIL-101(Cr)-EG | 710 | 0.47 | 0.43 | [25] |
| MIL-125(Ti) | 1160 | 0.47 | 0.36 | [26] |
| MIL-125-$NH_2$ | 830 | 0.35 | 0.37 | [25] |
| MIL-68(In) | 1100 | 0.42 | 0.32 | [25] |
| MIL-53(Al) | 1040 | 0.51 | 0.09 | [27] |
| MIL-53-$NH_2$(Al) | 940 | 0.37 | 0.09 | [25] |
| MIL-53-$(COOH)_2$-Fe | N.A. | N.A. | 0.16 | [25] |
| MOF-199(Cu) | 1430 | 0.72 | 0.55 | [25] |
| MOF-74(Co) | 1130 | 0.49 | 0.63 | [25] |
| MOF-74(Mg) | 1250 | 0.53 | 0.75 | [25] |
| MOF-74(Ni) | 1040 | 0.46 | 0.62 | [25] |
| MOF-801-P | 990 | 0.45 | 0.45 | [27] |
| MOF-801-SC | 690 | 0.27 | 0.35 | [27] |
| MOF-804 | 1145 | 0.46 | 0.290 | [27] |
| MOF-808 | 2060 | 0.84 | 0.735 | [22] |
| MOF-841 | 1390 | 0.53 | 0.640 | [22] |
| UIO-66 | 1290 | 0.49 | 0535 | [22] |
| UiO-66-$NH_2$ | 1328 | 0.70 | 0.38 | [22] |
| UiO-66-2,5-$(OMe)_2$ | 868 | 0.38 | 0.42 | [14] |
| UiO-67 | 2064 | 0.97 | 0.18 | [14] |
| Zn(NDI–SEt) | 888 | N.A. | 0.25 | [14] |
| Zn(NDI–SOEt) | 927 | N.A. | 0.30 | [14] |
| Zn(NDI–$SO_2$Et) | 764 | N.A. | 0.25 | [14] |

(continued)

**Table 3.1** (continued)

| MOF | BET surface area ($m^2/g$) | Pore volume ($cm^3/g$) | Water uptake capacity (g/g) | Literature |
|---|---|---|---|---|
| ZIF-8 | 1255 | 0.485 | 0.02 | [28] |
| $CO_2Cl_2$ (BTDD) | 1912 | 0.90 | 0.97 | [14] |
| Y-shp-MOF-5 | 1550 | 0.63 | 0.45 | [15] |

N.A. = Data not available

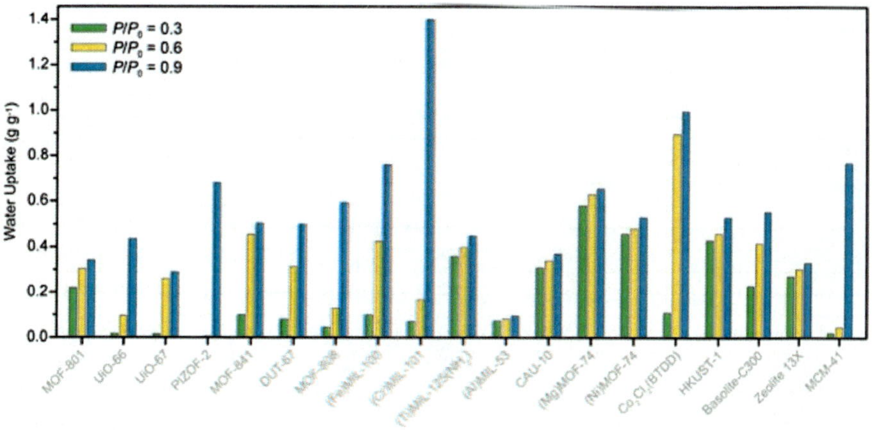


**Fig. 3.3** A bar diagram showing different water uptake capacities of well-known porous materials at RH values of 30, 60, and 90% [29]

(b) it must demonstrate repeatable performance under several cyclic adsorption-desorption tests, (c) it should possess excellent water uptake capacity within a narrow RH range, i.e., show a reversible Type V isotherm, and (d) it must be easily available, cheap and scalable for bulk production. It is worthy to note that only two MOFs: MIL-100(Fe) and CAU-10-H(Al) that exhibit these characteristics are commercially available and can be produced in bulk [20]. Besides, MIL-101(Cr), MOF-801(Zr), and UiO-66 are other well-known MOFs that demonstrate excellent potential in heat and moisture transfer applications (Fig. 3.3).

#### 3.2.2.1 MOFs-Based Desiccant Dehumidifiers

In one of the very first attempts, HKUST-1 MOF was coated on a fin-tube heat exchanger [30]. Since the isotherm shape of this MOF was characterized by Type-1 profile, its performance was compared to a silica gel-coated heat exchanger under dry and humid weather conditions. The material characterization results revealed that the MOF captured water vapor significantly higher than silica gel in low-to-medium humidity ranges. In contrast, silica gel demonstrated up to 40% higher uptak

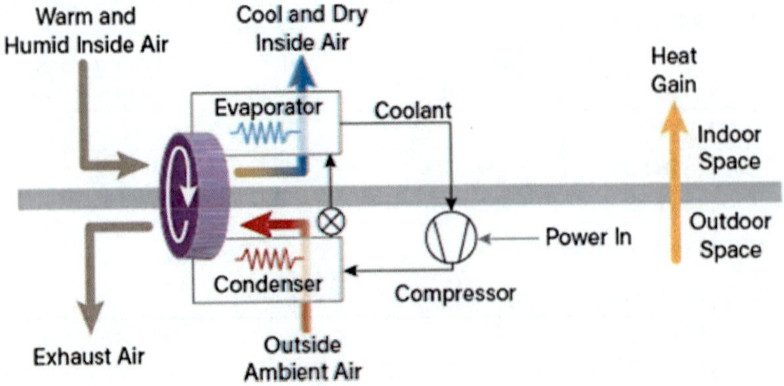


**Fig. 3.4** A MOF-based cartridge (purple) coupled to an existing air conditioner [32]

capacity under high humidity conditions. While a regeneration temperature of 60 °C was deemed to be sufficient for HKUST-1, silica gel's performance deteriorated by almost 2 times. Later, Zu and co-workers [31] retrofitted a heat pump unit with a MIL-100(Fe) heat exchanger and compared the performance with silica gel and zeolite coated heat exchangers. This DCHE could be regenerated at approximately 50 °C and recorded almost 36% savings in electrical power consumption. In addition, with an aim to develop a sustainable air-conditioner, a startup from the Massachusetts Institute of Technology (MIT) developed a desiccant wheel-based MOF dehumidifier and integrated with a direct expansion MVC system, as shown in Fig. 3.4. The MOF-impregnated wheel dehumidifies the air before the evaporator and the air heated by the condenser is used for regenerating the wheel [32].

MIL-53(Cr) and Zn(BDC)-(TED) have shown great potential when applied to heat pumps/air-conditioners [33]. However, their smaller pore size and hysteresis limit their practical usage. Generally, MOFs with larger pores are preferred since they lead to a stronger water-water molecule interaction. The heat released during the sorption process in such materials approaches water's latent heat of vaporization, thereby lowering the desorption temperatures [34]. However, a larger pore size promotes capillary condensation, and the possibility of hysteresis increases significantly. Therefore, a tradeoff arises while tailoring the MOFs structure, and the parameters must be adjusted to achieve maximum performance. ZJNU-30 was identified for possible application in coolers as it achieved a high cooling capacity of 550kWh/$m^3$ and recorded around 1.2 g/g isothermal capacity, which is around 4.5 times higher than silica gel [35]. MOF-74(Ni) with extended phenyl rings were examined for adsorption performance with water and R134a adsorbates. The experimental results demonstrated excellent adsorbing characteristics of the MOF with 0.9 g/g of water and about 0.8 g/g of R134a [36].

#### 3.2.2.2 Future Potential of MOFs

Although an extensive list is already available in the literature to guide the selection of MOFs for possible heat transformation applications, they face a tough challenge competing with conventional desiccants. This is because of the substantial capital investment required in MOF production and difficulties in achieving scalability. As the starting materials are expensive, efforts to synthesize MOFs from low-cost synthons are imperative. Further, to coat MOFs on fin-tube heat exchangers, an appropriate binder selection process, as described in Sect. 2.4, is needed for achieving enhanced thermal and mass transport rates. Notably, the compatibility between MOFs and binders must be carefully monitored to prevent the blockage of pores of MOFs. When these factors are accounted for suitably, MOFs are expected to be game-changers in energy-related applications. These advanced applications are covered at a greater depth in Chap. 5.

## 3.3 Desiccant Synthesis and Characterization

To facilitate the application of a specific desiccant, sorption and desorption characteristics of the desiccant must be understood and analyzed. These results are readily accessible for commercial desiccants like silica gel and different types of zeolites/molecular sieves. However, comprehensive characterization results are seldom available for composite ceramic/inorganic desiccants or advanced materials like superabsorbent polymers and MOFs. Table 3.2 presents the list of experimental and theoretical techniques that can be used for characterizing a desiccant material. The experimental results from the material characterization studies act as inputs in formulating theoretical performance prediction approaches for desiccant dehumidifiers. The quintessential desiccant characteristics: isotherms and kinetics, required to support the governing principles of mass and energy conservation, are presented in the ensuing section. While this characterization analysis is carried out for silica gel and composite polymer desiccants, one can apply the approach and conduct similar studies even for MOFs.

### *3.3.1 Development of Composite Superabsorbent Desiccant Solutions*

Silica gel (RD 780), anhydrous LiCl, $HCO_2K$, PVA, and sodium polyacrylate-based SAP powders have been used to prepare desiccant coating solutions. The thermo-physical properties of silica gel, PVA, SAP, LiCl, and $HCO_2K$ used in this study are listed in Table 3.3 [47–50]. Table 3.4 lists the steps involved in preparing the desiccant solution that is suitable for coating on metallic surfaces or heat exchangers of complex shapes and sizes.

**Table 3.2** List of experimental and theoretical techniques available for measuring the characteristic properties of desiccants

| Desiccant's characteristic properties | Experimental/computational techniques |
|---|---|
| Pore characteristics: Pore volume, radius, BET surface area | $N_2$ sorption at 77 K [37] |
| • Structural integration of the desiccant<br>• Structural stability | Powder X-ray diffraction (PXRD) [38] |
| • Surface morphology<br>• Bulk properties of crystalline materials upon sorption | • Field emission scanning electron microscopy (FESEM) [39]<br>• Transmission electron microscopy (TEM) [40] |
| In situ surface topology | Atomic force microscopy (AFM) [41] |
| Locating water interaction sites in the desiccant | Extended X-ray absorption fine structure (EXAFS) [42] |
| In situ water interaction with the desiccant | • Infrared (IR)<br>• Raman spectroscopic techniques |
| Internal defects in desiccant structure | Confocal Microscopy |
| Water sorption isotherms | • Volumetric method [43]<br>• Gravimetric method [44] |
| Water sorption kinetics, mass diffusivity, and activation energy | • Gravimetric method<br>• Concentration swing frequency response (CSFR) apparatus |
| Heat of sorption/desorption and heat capacity | Thermogravimetric analysis |
| Thermal conductivity | Steady-state or transient thermal conductivity analyzers [45] |
| • Computational characteristics measurement<br>• Explaining the water-desiccant interactions<br>• Calculation of adsorption enthalpy | • Molecular dynamics (MD) [46]<br>• Grand Canonical Monte-Carlo (GCMC) [28] |

**Table 3.3** Thermo-physical properties of Silica gel, PVA, SAP, LiCl, and $HCO_2K$

| Properties | Silica gel | PVA | SAP | LiCl | $HCO_2K$ |
|---|---|---|---|---|---|
| Molecular Weight, g/mol | 60.084 | 44.05 | 94.04 | 42.4 | 130.14 |
| Density, g/cm$^3$ | 0.7-0.8 | 1.521 | 1.1–1.4 | 2.07 | 1.91 |
| Solubility (in $H_2O$) | Poorly soluble. Forms a suspension | At high temperature | Good | Very good | Very good |
| Specific heat capacity, kJkg$^{-1}$K$^{-1}$ | 0.921 | 1.4 | 0.96 | 1.132 | – |
| Toxicity | Low | Low | Low | Low | Low |

**Table 3.4** Steps involved in desiccant solution preparation

| Desiccant solution | Method |
|---|---|
| Silica gel | • A 5 w% silica gel suspension (i.e., 5 g of silica gel in 100 ml of water) is made by adding silica gel to distilled water at room temperature<br>• A continuous stirring ensures that the suspension remains homogeneous and suitable for coating<br>• HEC powder with 3.3 w% concentration is added to the silica gel solution to achieve suitable binding of silica gel on the Al fin [43] |
| PVA/LiCl | • A 2 w% PVA solution is prepared by adding PVA into distilled water at room temperature and continuous stirring via a magnetic stirrer at 50–70 °C until a homogeneous solution is obtained<br>• Appropriate amounts of LiCl are added to produce PVA/LiCl solutions with different LiCl concentrations. A 0 w% PVA/LiCl solution indicates that the solution is pure PVA solution whereas 100 w% LiCl implies pure LiCl solution. The combinations between these extremes can be represented as 16.7, 33.3, 50, and 66.7 w% LiCl concentrations |
| SAP/LiCl | • A 2 w% SAP solution is prepared by adding powdered SAP into distilled water. The mixture is stirred using a magnetic stirrer at 50–70 °C until a homogeneous solution is obtained<br>• Like PVA/LiCl desiccants, by controlling the mass of LiCl, desiccant solutions with multiple salt concentrations of LiCl are produced, namely, 0, 16.7, 33.3, and 50 w% |
| SAP/$HCO_2K$ | • Preparing composite PVA desiccants with $HCO_2K$ results in a heterogeneous mixture [51]. As a result, it is deemed to be unsuitable for coating<br>• SAP and $HCO_2K$ powders are added in distilled water and continuously stirred at 60 °C to produce SAP/$HCO_2K$ solutions with four different concentrations (16.7, 33.3, 50, and 75 w%) of $HCO_2K$ |

## 3.3.2 *Isotherms*

As mentioned earlier in Sect. 2.1, isotherms convey the desiccant's equilibrium moisture sorption capacity at a particular temperature and RH. The isotherms can be obtained via gravimetric measurements in a controlled chamber. In this technique, the mass of the desiccant is directly weighed with a highly accurate mass balance. The chamber's temperature is maintained at a constant value, and its RH is raised stepwise from 0 to 95%. The desiccant's instantaneous water uptake ($q_t$), which represents the dynamic change of desiccant's mass, is then computed using Eq. (3.1). $q_t$ is measured continuously until its change with respect to time is infinitesimal (i.e., less than 0.001% continuously for 3–4 min). At this stage, the desiccant is said to have achieved its equilibrium, and the value of $q_t$ corresponds to the desiccant's equilibrium uptake capacity ($q_e$).

$$q_t = \frac{m_w - m_d}{m_d} \tag{3.1}$$

where $m_w$ is the mass of the wet desiccant during the sorption/desorption process; and $m_d$ is the mass of the dry desiccant.

#### 3.3.2.1 Isotherm Results for Composite Polymer Desiccants

Figure 3.5 illustrates the equilibrium sorption capacity (represented as a percentage of the desiccant's initial dry mass) of silica gel and multiple concentrations of PVA/LiCl desiccants. The isotherm shape of silica gel corresponds to the Type-I IUPAC classification, and its highest sorption capacity is less than 30%. Since the moisture capturing process is fundamentally limited to monolayer adsorption, silica gel's equilibrium sorption capacity remains markedly limited. In contrast, the sorption capacity of the composite PVA desiccant is capable of reaching up to 300% of the desiccant's initial mass, as presented in Fig. 3.5b. Such a marked rise in adsorption capacity is due to the combination of several water sorption mechanisms as opposed to monolayer adsorption in silica gel.

The LiCl molecules in the composite desiccant attract water due to their excellent hygroscopicity and dissociate into $Li^+$ and $Cl^-$ ions. The presence of water molecules in the composite desiccant promotes swelling of the polymer chain and increases the polymer chain's interstitial volume. Consequently, the room for storing water vapor in the desiccant is raised significantly, and the desiccant absorbs more water vapor [52]. Like the observations made in Sect. 2.3, higher LiCl concentration enhances the composite desiccant's sorption capacity and promotes deliquescence. Since deliquescence is undesired, the quantity of LiCl in the composite desiccant must be restricted appropriately.

Figure 3.6 shows the states of two PVA/LiCl desiccants with 50 and 66.7 w% LiCl after sorbing water vapor. The 50 w% configuration remains intact with the substrate, while the 66.7 w% liquifies upon capturing moisture. Therefore, the maximum limit of LiCl in the composite PVA desiccant is restricted to 50 w% to realize improvements

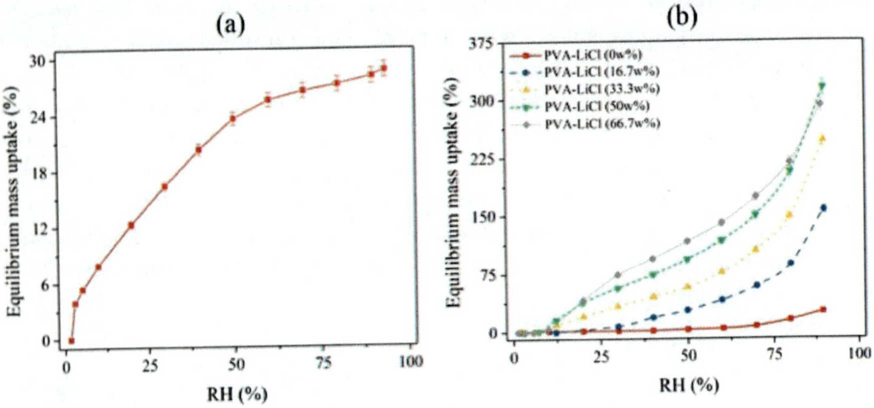


**Fig. 3.5** Water vapor sorption isotherms at 30 °C for **a** silica gel and **b** PVA with 0–66.7 w% LiCl

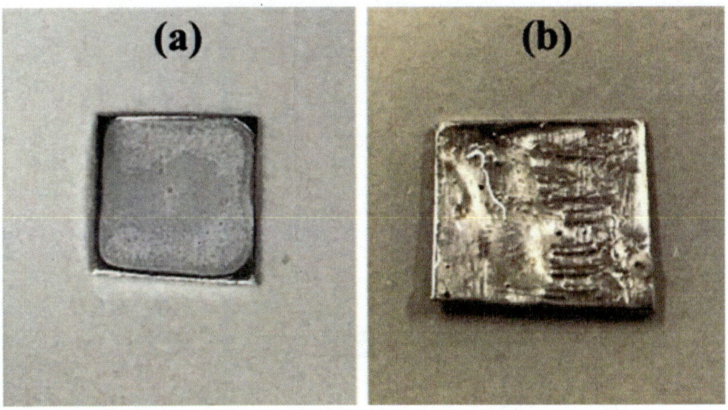


**Fig. 3.6** Photographs of **a** PVA-LiCl (50 w%) and **b** PVA-LiCl (66.7 w%) coated Al sheet after moisture sorption at 80% RH

in sorption capacity and desiccant's coating stability. At 80% RH, PVA-LiCl (50 w%)'s equilibrium capacity is 177.2%, an ultra-high-capacity compared to silica gel's 28%. This marked increase in adsorption capacity and easy coating ability paves the path for the composite polymer desiccants as next-generation dehumidification materials.

#### 3.3.2.2 Heat of Sorption

The moisture sorption process by the desiccant is exothermic. The sorption heat describes the ratio of heat released during the sorption process to the amount of water vapor captured by the desiccant. It is worthy to note that the heat of sorption is not a material-specific property. Its value depends on the reaction condition in which sorption/desorption processes occur. Also, the fundamental property impacting the heat of sorption is the RH. For the commonly encountered environmental temperatures and pressures in most desiccant dehumidification applications, the heat of sorption can be treated as a function of humidity alone [53]. It can be directly measured by conducting calorimetric measurements simultaneously during the sorption/desorption process via thermo-gravimetric analysis (TGA). On the other hand, it can also be indirectly obtained by conducting the gravimetric measurements of isotherms at different temperatures. In the indirect approach, the Clausius Clapeyron equation [54], shown in Eq. (3.2), is employed to derive the sorption heat ($\Delta H_s$).

$$\ln(P_w) = -\frac{\Delta H_s}{R}\frac{1}{T} + c \tag{3.2}$$

where $P_w$ is the partial pressure of water vapor; $R$ is the ideal gas constant; $T$ is the absolute temperature; and $c$ is a constant.

In Fig. 3.7a and b, isotherms of silica gel and SAP-LiCl (50 w%) are presented in the range spanning 30–40 °C. Their equilibrium isothermal capacity is not affected

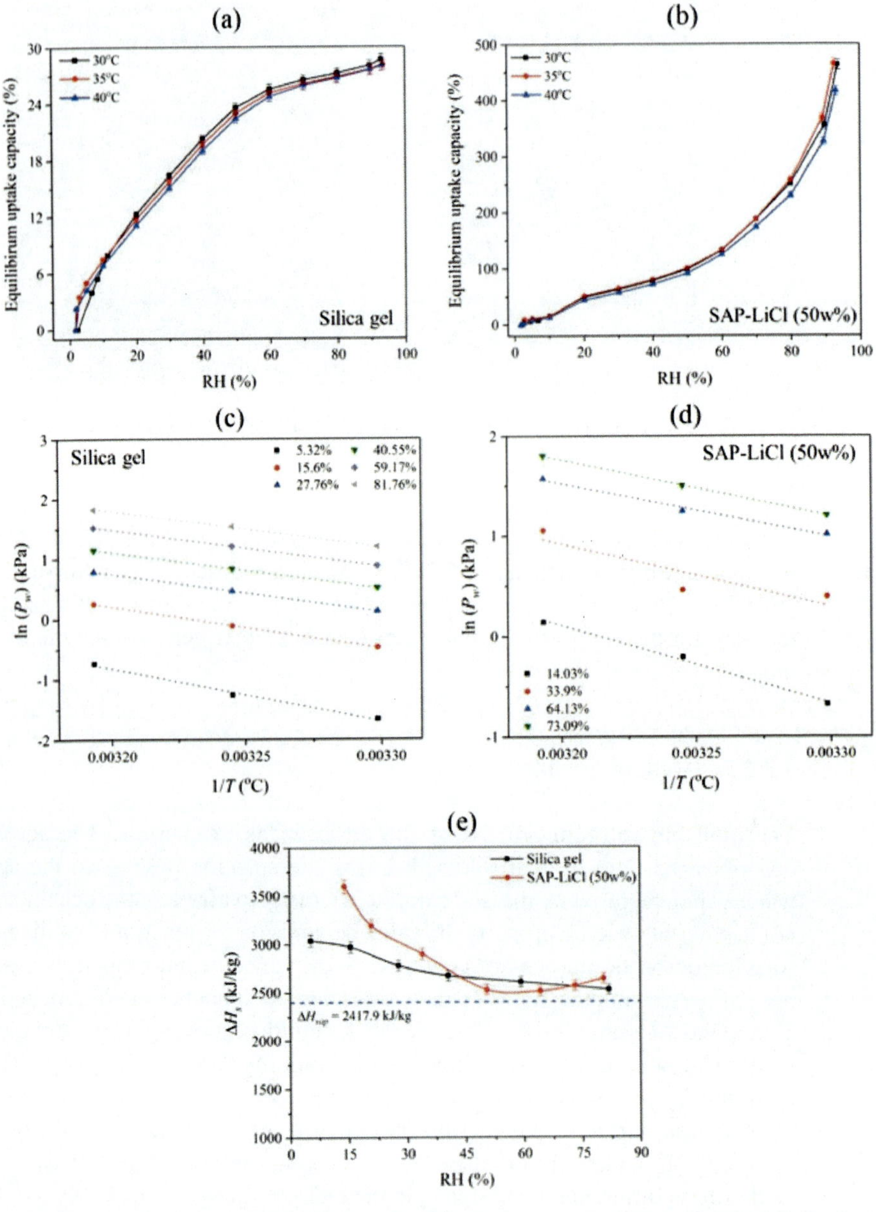


**Fig. 3.7** **a–b** Isotherms of silica gel and SAP-LiCl (50 w%) obtained between 30 and 40 °C; **c–d** ln ($P_w$) versus 1/$T$ plot obtained from the isotherms of silica gel and SAP-LiCl (50 w%); and **e** a comparison of heat of sorption of silica gel and SAP-LiCl (50 w%) with the latent heat of vaporization of water $\left(\Delta H_{vap}\right)$

by temperature, and it varies with RH alone. To employ the Clausius Clapeyron equation to the desiccant's isotherms, the natural logarithm of $P_w$ is first computed, and the slopes of ln ($P_w$) versus $1/T$ graphs are then plotted, as shown in Fig. 3.7c and d. The results of heat of sorption ($\Delta H_s$) are shown in Fig. 3.7e. The latent heat of vaporization of water $\left(\Delta H_{vap}\right)$ at 35 °C and atmospheric pressure is also specified. For both the desiccant types, $\Delta H_s$ is greater than $\Delta H_{\text{vap}}$ at lower RH, and it gradually approaches $\Delta H_{\text{vap}}$ at higher RH values. This phenomenon is attributed to the interaction of the desiccant with water vapor molecules. The sorption heat is high at lower ranges of RH since a strong molecular interaction exists between desiccant and water. When the desiccant absorbs a significant quantity of water vapor, the water-to-water interaction dominates the desiccant-to-water interaction. This results in $\Delta H_s$ being close to $\Delta H_{\text{vap}}$. Lastly, although SAP-LiCl (50 w%) captures approximately 12 times higher moisture from air than silica gel, its average heat of sorption $\left(\Delta \bar{H}_s = 2842 \text{ kJ/kg}\right)$ is only 3% higher than silica gel $\left(\Delta \bar{H}_s = 2764 \text{ kJ/kg}\right)$.

### *3.3.3 Kinetics*

Water sorption/desorption kinetics determine the rate at which the desiccant attains its equilibrium capacity at a particular temperature and RH. It depends on the geometric characteristics of the desiccant, such as its surface area and thickness. When the kinetics of different desiccants need to be compared, their geometrical aspects need to be identical. To evaluate the desiccant's kinetics, the instantaneous water uptake profile, obtained via Eq. (3.1), is analyzed. The gravimeter is maintained between 20 and 40 °C at a constant RH of 80% during sorption and between 40 and 80 °C at 0% RH during desorption processes.

#### 3.3.3.1 The Linear Driving Force Approximation

The desiccant's kinetics profile can be mathematically represented using semi-empirical models. Such representations produce valuable desiccant characterization data that can be integrated with the energy/mass conservation laws for desiccant dehumidifiers. Many semi-empirical models, namely, linear driving force (LDF) approximation, semi-infinite model, Fickian diffusion, and second-order approximation, have been developed to represent the desiccant's kinetics [52, 55, 56]. However, the LDF approximation is usually preferred to regress the kinetics data as this method has proven to be consistent and is straightforward. In the LDF approximation, the desiccant's instantaneous water vapor sorption/desorption rate is represented via Eq. (3.3). When this equation is integrated and appropriate limits applicable for sorption/desorption processes are introduced, Eqs. (3.4) and (3.5) are obtained.

**Table 3.5** Regression results obtained by using LDF approximation

| Desiccant | Nature | $q_0$ (kg/kg) | $k$ (1/s) | $q_e$ (kg/kg) | $R^2$ | NRMSE (%) |
|---|---|---|---|---|---|---|
| Silica gel | Adsorption | 0.03 | $2.36\times10^{-3}$ | 0.26 | 0.9942 | 2 |
| | Desorption | 0.27 | $3.42\times10^{-3}$ | $5.17\times10^{-5}$ | 0.9842 | 4 |
| PVA-LiCl (50 w%) | Sorption | 0.15 | $1.09\times10^{-4}$ | 1.78 | 0.9866 | 2 |
| | Desorption | 1.78 | $5.93\times10^{-4}$ | 0.22 | 0.9988 | 3 |

$$\frac{dq_t}{dt} = k(q_e - q_t) \tag{3.3}$$

$$q_t = q_e - \exp\{\ln(q_e - q_0) - kt\} \tag{3.4}$$

$$q_t = q_e + \exp\{\ln(q_0 - q_e) - kt\} \tag{3.5}$$

where $q_t$ and $q_e$ are the desiccant's instantaneous and equilibrium water vapor quantity, $q_0$ is the desiccant's initial water content present in desiccant, and $k$ is the kinetic constant.

The regression results of the kinetics data are presented in Table 3.5 and Fig. 3.8. The coefficient of determination ($R^2$) is close to 1 for sorption/desorption kinetics of silica gel and PVA-LiCl (50 w%) coated Al sheets. The normalized root mean square error (NRMSE) ranges spanning 2–4%. The excellent regression result shows that the LDF approximation is sufficient to portray silica gel and composite polymer desiccant's sorption/desorption rates. Figure 3.9 compares the sorption kinetics of silica gel and PVA-LiCl (50 w%) coated Al sheets at 30 °C and 80% RH. While the pores in silica gel are filled at around 2500 s, PVA-LiCl (50 w%) superabsorbent continues to capture water vapor at over 10,000 s. At the point of time that the silica gel becomes saturated, PVA-LiCl (50 w%) is able to capture around 57% of water vapor, two times that of silica gel. This observation indicates that the composite superabsorbent desiccant's higher capacity to absorb moisture promotes its reaction rate.

#### 3.3.3.2 Effect of Salt Concentration

The experimental results of the composite PVA desiccants' sorption and desorption kinetics are depicted in Fig. 3.10. It is apparent that a higher LiCl concentration promotes the sorption and desorption rates, and the respective regression results are presented in Fig. 3.11a and b. The reason for the enhanced sorption rate is primarily attributed to the increase in equilibrium capacity. As the equilibrium sorption capacity improves, it promotes the driving force for moisture sorption and enhances the desiccant's reaction rate. Contrastingly, the kinetics constant's variation with respect to the LiCl concentration remains insignificant. Likewise, due to the higher initial water

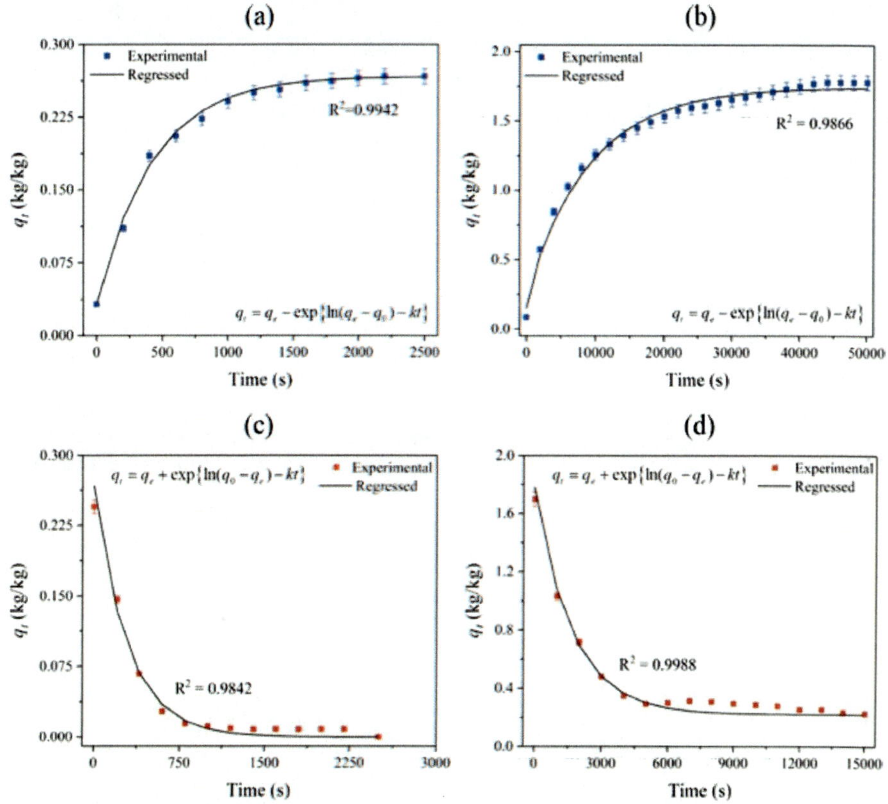


**Fig. 3.8** Experimental and regression results of sorption kinetics of **a** silica gel and **b** PVA-LiCl (50 w%) at 30 °C and 80% RH; and experimental and regression results of desorption kinetics of **c** silica gel and **d** PVA-LiCl (50 w%) at 60 °C and 0% RH

**Fig. 3.9** Sorption kinetics comparison of silica gel and PVA-LiCl (50 w%) at 30 °C and 80% RH

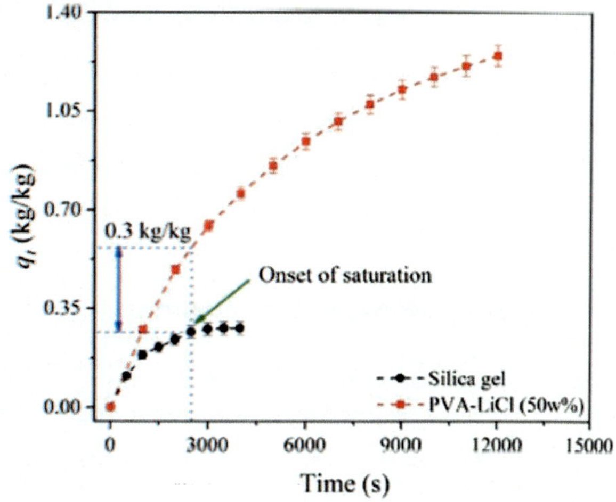

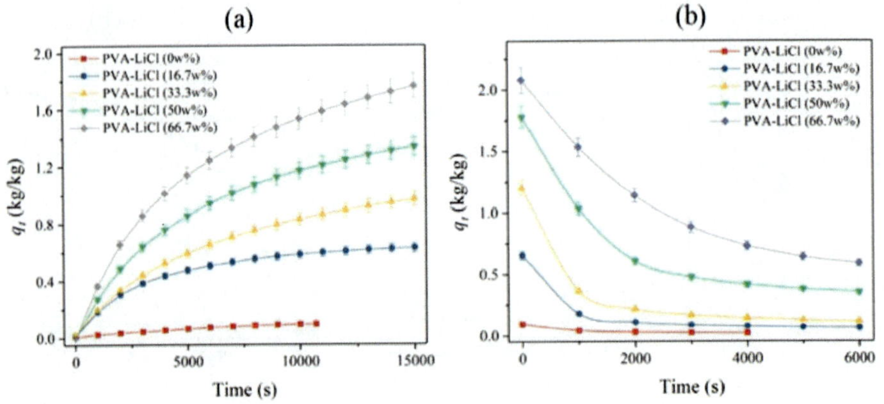


**Fig. 3.10** Experimental result of composite PVA with increasing concentration of LiCl from 0 to 66.7 w% for **a** sorption kinetics at 30 °C and 80% RH and **b** desorption kinetics at 60 °C and 0% RH

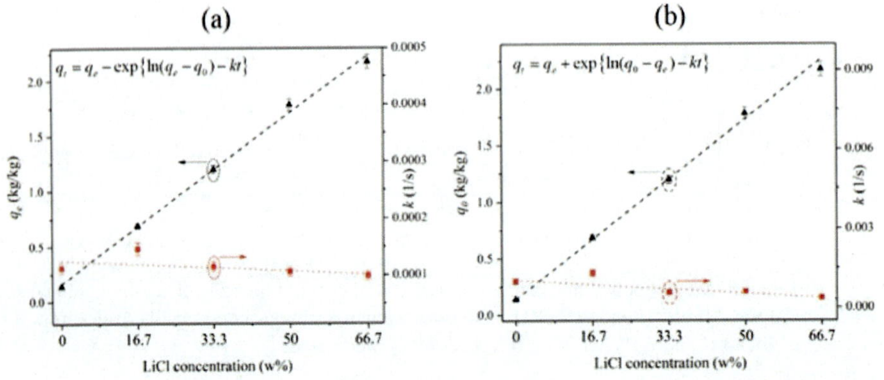


**Fig. 3.11** **a** Variation of equilibrium uptake capacity and kinetic constant for sorption process at 30 °C and 80%RH; and **b** variation of initial water content and kinetic constant for desorption process at 60 °C and 0%RH for composite PVA with increasing LiCl concentration from 0 to 66.7 w%

concentration in the desiccant, the desorption rate is significantly enhanced when the amount of LiCl concentration increases.

#### 3.3.3.3 Effect of Temperature

The experimental findings of increasing reaction temperature on desiccant kinetics are shown in Fig. 3.12. The regression data for the equilibrium capacity and the kinetic constant are illustrated in Fig. 3.13. Since the kinetic constants improve at higher temperatures, the sorption and desorption rates are enhanced. On the other hand, the equilibrium capacity remains relatively unchanged as it varies only with respect to

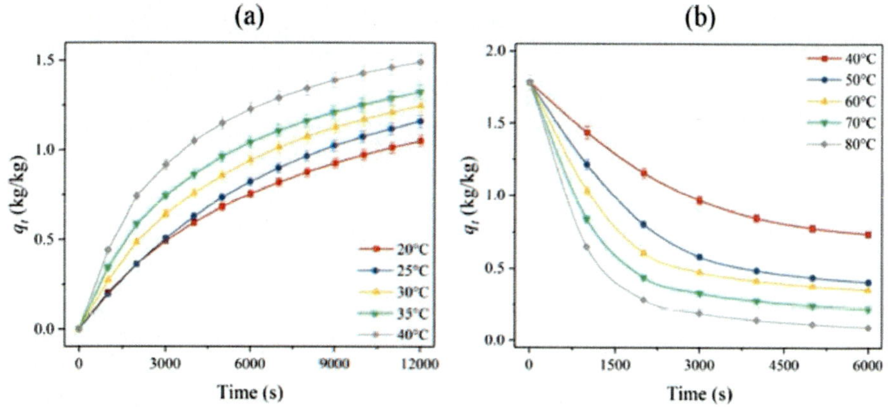


**Fig. 3.12** Experimental result of PVA-LiCl (50 w%) for **a** sorption kinetics with temperature ranging from 20 to 40 °C at 80% RH and **b** desorption kinetics with temperature ranging from 40 to 80 °C at 0% RH

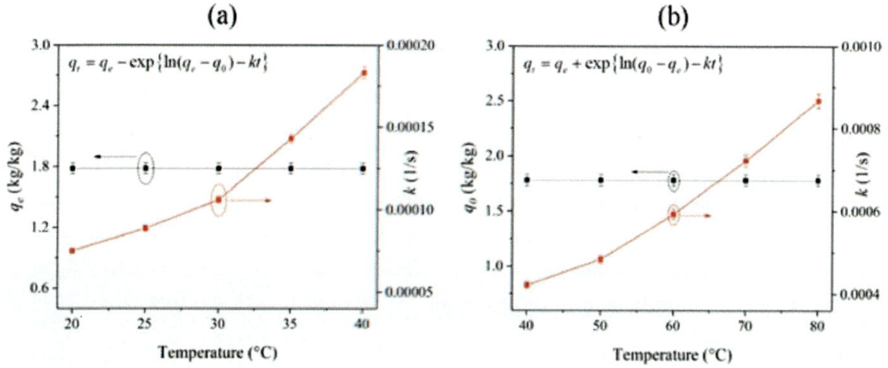


**Fig. 3.13** **a** Variation of equilibrium uptake capacity and kinetic constant between 20 and 40 °C during sorption process at 80% RH, and **b** variation of initial water content and kinetic constant between 40 and 80 °C and 0% RH during desorption process of PVA-LiCl (50 w%)

RH. Additionally, the desorption kinetic constant is an order of magnitude greater than the sorption kinetic constant. This is because of higher reaction temperatures typically adopted during the regeneration process. As a result, the regeneration time would be much lower than the time required for dehumidification and not act as a limiting factor.

### 3.3.4 Activation Energy

The kinetic constant strongly depends on the reaction temperature, which can further be validated by correlating its dependence via the Arrhenius equation, as shown in Eq. (3.6).

$$k = A\exp\left(-\frac{E_a}{RT}\right) \tag{3.6}$$

where, $E_a$ is the activation energy, $A$ is the pre-exponential factor, $R$ is the universal gas constant, and $T$ is the absolute temperature. Figure 3.14 portrays the Arrhenius plot, while Table 3.6 lists the values of sorption/desorption activation energy and pre-exponential factors. Due to the higher operating temperatures commonly adopted during regeneration, the desorption process exerts less inertia for water vapor migration. As a result, the desorption activation energy is about 2 times lower than that of sorption activation energy.

## 3.4 Experimental Performance Evaluation of DCHEs

In the previous section, the results on the material characterization of different composite polymer desiccants were presented. This section specifically focuses on evaluating the dynamic dehumidification and thermal performance of DCHEs. The impact of composite polymer DCHEs on moisture removal capacity and process efficacy is studied, and their results are benchmarked against silica gel-coated heat exchangers.

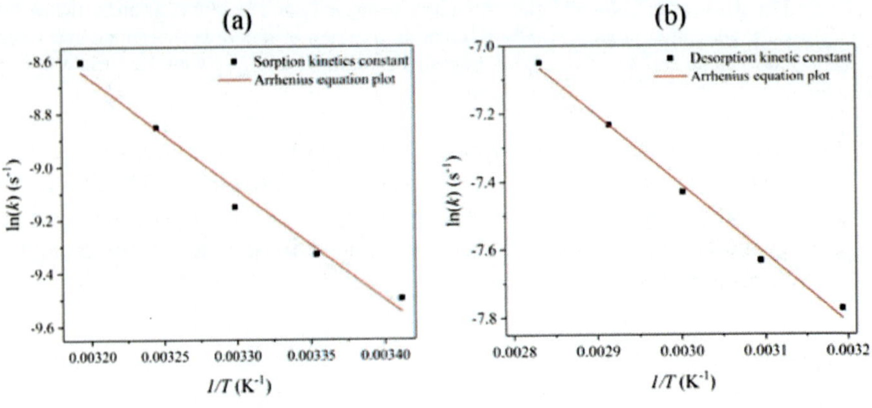


**Fig. 3.14** Arrhenius plot of PVA-LiCl (50 w%) coated Al sheet during **a** sorption and **b** desorption processes

**Table 3.6** Results of activation energy and pre-exponential factor of PVA-LiCl (50 w%) coated Al sheet during sorption and desorption processes

| $T$(°C) | $k$ (1/s) | $1/T$ ($K^{-1}$) | ln($k$) (1/s) | $E_a$ (kJ/mol) | $A$ (1/s) |
|---|---|---|---|---|---|
| *PVA-LiCl (50 w%) Sorption (RH = 80%)* | | | | | |
| 20 | $7.50 \times 10^{-5}$ | 0.003411 | −9.49765 | 34.43 | 97.39 |
| 25 | $8.87 \times 10^{-5}$ | 0.003354 | −9.33063 | | |
| 30 | $1.06 \times 10^{-4}$ | 0.003299 | −9.15207 | | |
| 35 | $1.43 \times 10^{-4}$ | 0.003245 | −8.85267 | | |
| 40 | $1.83 \times 10^{-4}$ | 0.003193 | −8.60641 | | |
| *PVA-LiCl (50 w%) Desorption (RH = 0%)* | | | | | |
| 40 | $4.20 \times 10^{-4}$ | 0.003193 | −7.77442 | 16.97 | 0.278 |
| 50 | $4.84 \times 10^{-4}$ | 0.003095 | −7.63354 | | |
| 60 | $5.93 \times 10^{-4}$ | 0.003002 | −7.43093 | | |
| 70 | $7.22 \times 10^{-4}$ | 0.002914 | −7.23319 | | |
| 80 | $8.67 \times 10^{-4}$ | 0.002832 | −7.04991 | | |

## 3.4.1 Experimental Facilities

### 3.4.1.1 Integrated with a Humidity Chamber

The testing facility for evaluating the performance of a DCHE with a single outdoor air supply is shown in Fig. 3.15. A humidity cum temperature chamber is employed to achieve the desired climatic conditions. An air blower is installed to regulate the flow rate of air. Two resistance temperature detectors (RTDs) are placed at the inlet and outlet of the testing chamber to measure the dry and wet bulb temperatures. Air velocity transmitters are positioned before the inlet and after the chamber's outlet. The face velocity and mass flow rate of the air passing through the DCHEs can be computed via Eqs. (3.7) and (3.8), respectively. A variable-area flowmeter with ± 4% full-scale accuracy is used to record the water flow rate. LabVIEW 2015 data acquisition system is employed to record the experimental data. A summary of the measuring instruments used in the setup along with their accuracies is listed in Table 3.7. The humidity chamber and the air blower are first set to the required inlet conditions. The air valves are adjusted such that the air passes through the by-pass line. Once the stable inlet air conditions are obtained, the air interacts with the DCHE. Concurrently, cooling water flowing in the tubes takes away the heat released during the desiccant's sorption process. For regeneration, the cooling water flow is switched to the hot water flow.

$$V_f = \frac{\pi \left(d_{pipe}\right)^2 V_{sen}}{4LW} \tag{3.7}$$

$$\dot{m}_a = 3600 \rho_a V_f L W \tag{3.8}$$

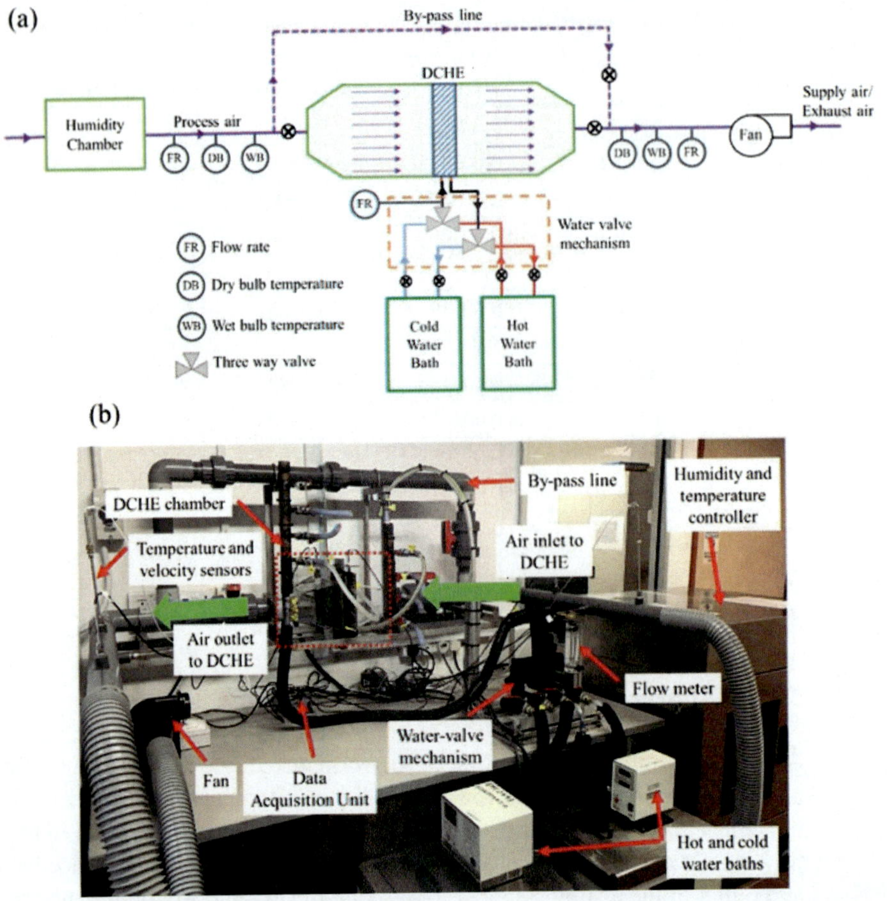


**Fig. 3.15** **a** A schematic and **b** a photograph of the DCHE dynamic performance testing experimental setup

**Table 3.7** Specifications of the sensors and measuring instruments used in the DCHE performance testing setup

| Parameter | Sensor/instrument | Range | Accuracy |
|---|---|---|---|
| Temperature | RTD temperature sensor | −29–100 °C | 1/10 DIN |
| Airflow rate | Air velocity meter | 0–10 m/s | ±2.5% (Full Scale) |
| Water flow rate | Variable area flowmeter | 1–7 L/min | ±4% (Full Scale) |

where $V_f$ is the face velocity of air at the heat exchanger inlet; $d_{pipe}$ is the diameter of the pipe where the air velocity meter is mounted; $V_{sen}$ is the velocity of air measured by the air velocity meter; $L$ and $W$ are the respective length and width of the heat exchanger; $\dot{m}_a$ is the mass flow rate of air; and $\rho_a$ is the density of air.

#### 3.4.1.2 Provisions for Indoor and Outdoor Air Mixing

Figure 3.16 illustrates another experimental system that can be employed to evaluate the DCHE's dynamic performance. Unlike the previous setup, which consists of a temperature and humidity chamber to provide the necessary inlet air conditions, this test facility uses outdoor and indoor air. To achieve stable temperature and humidity

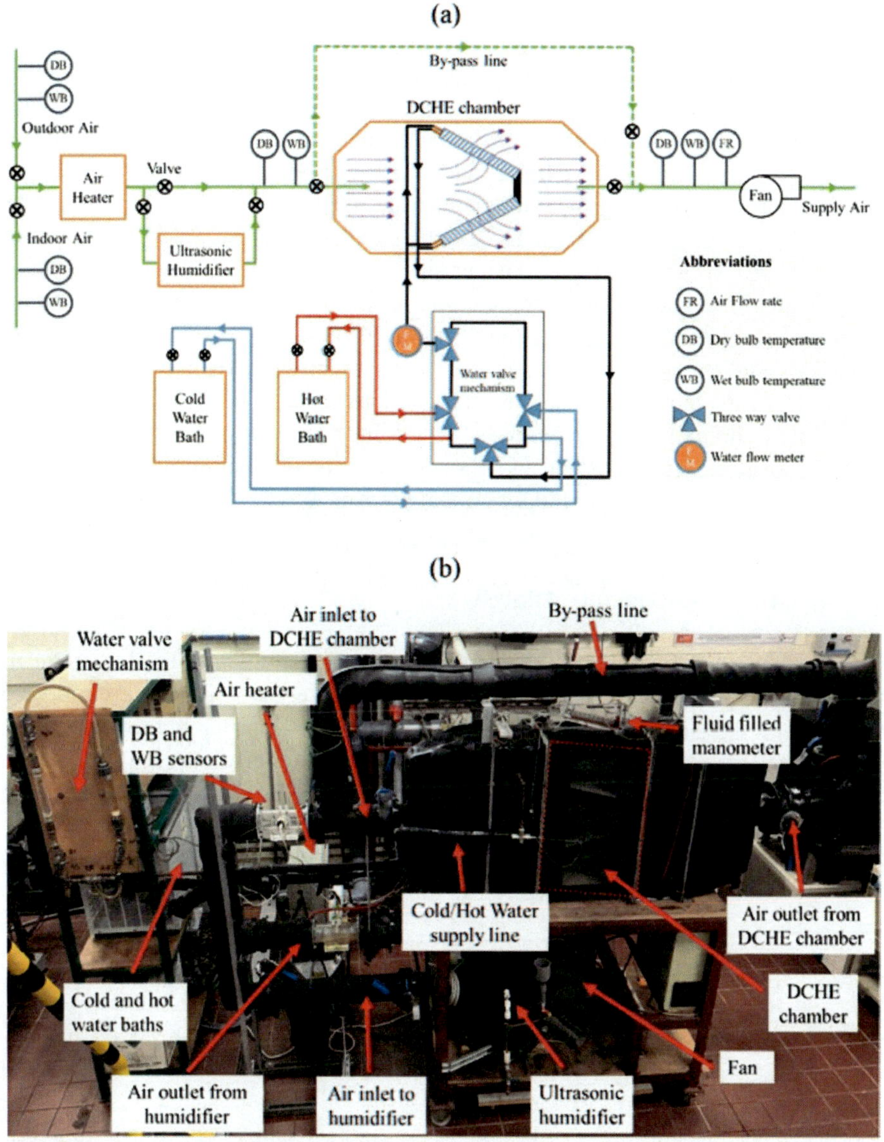


**Fig. 3.16** **a** A schematic and **b** a photograph of the testing facility of DCHE

**Table 3.8** Specifications of the sensors and measurement instruments used in the DCHE performance testing setup

| Parameter | Sensor/instrument | Range | Accuracy |
|---|---|---|---|
| Temperature | Thermistors | −10–100 °C | ±0.15 °C |
| Airflow rate | Thermal mass flow meter | −10–100 °C | ±0.5% (Full Scale) |
| Water flow rate | Compact vortex flowmeter | 3.2–22 L/min | ±2.5% (Full Scale) |
| Pressure drop | Fluid-filled manometer | 0–50 mm $H_2O$ | ±1.0% (Full Scale) |

conditions, an auxiliary air heater and an ultrasonic humidifer are included. Two DCHEs are installed at an tilted angle of 30° to achieve laminar flow and improve the interaction area between the desiccant and moist air. Dry bulb and wet bulb thermistors are attached before and after the DCHE chamber and moist air properties are recorded. The airflow rate is measured using a thermal mass flow meter, and it records the flow rate at the DCHE chamber's outlet. The water flow rate is measured using a vortex flow meter. The air-side pressure drop ($\Delta P$) is obtained using a fluid-filled manometer. The specifications and accuracy of the measuring instruments are listed in Table 3.8. During the dehumidification process, the outdoor air first passes through the air heater and ultrasonic humidifier to achieve the desired inlet conditions. Until the temperature and humidity values reach a steady state, the air is made to pass through the by-pass line. The air interfaces with the dry DCHEs to facilitate moisture sorption. When switching occurs, the outdoor air and cooling water flows are interchanged with indoor air and hot water.

#### 3.4.1.3 Uncertainty Analysis

The values recorded by the sensors or instruments fall within a certain range due to the uncertainties associated with the measurements. The uncertainties associated with the directly measured parameters are readily available from the accuracy data of the instruments. However, the error propagation method [57] (Eq. 3.9) is applied to compute the uncertainties of parameters derived from the directly measured ones.

$$\delta y = \sqrt{\left[\sum_{i=1}^{n}\left(\frac{\partial f}{\partial x_i}\right)^2 (\delta x_i)^2\right]} \tag{3.9}$$

where $\delta y$ is the uncertainty associated with the derived parameters; $\delta x$ is the uncertainty associated with the measured parameters; $n$ is the number of parameters directly measured using the sensors; and $f$ represents the mathematical relationship between the derived and measured parameters.

For example, the air's dry bulb ($T_{db}$) and wet bulb ($T_{wb}$) temperatures are directly measured by the thermistors/resistance temperature detectors (RTDs). When the humidity ratio of air ($\omega$) needs to be evaluated from the temperatures measured, the uncertainty associated with $\omega$ is indirectly evaluated by computing the direct uncertainties from the measured parameters. Buck [58], Modified Ferrel [59], and the ideal gas equations are employed for deriving $\omega$, as shown in Eqs. (3.10)–(3.12).

$$P'_{sat} = 0.61121 \exp\left\{\left(18.678 - \frac{T_{wb}}{234.5}\right)\left(\frac{T_{wb}}{257.14 + T_{wb}}\right)\right\} \tag{3.10}$$

$$P_v = P'_{sat} - 0.00066 P_{atm}(T_{db} - T_{wb})\left(1 + \frac{1.8}{1571}T_{db}\right) \tag{3.11}$$

$$\omega = 0.622\frac{P_v}{P_{atm} - P_v} \tag{3.12}$$

where $P'_{sat}$ is the saturated vapor pressure of water vapor at $T_{wb}$; $P_v$ is the vapor pressure of water in moist air; and $P_{atm}$ is the atmospheric pressure.

The uncertainty in the humidity ratio ($\delta\omega$) is computed via Eq. (3.9) and is as shown in Eq. (3.13) and the percentage error is computed from Eq. (3.14) and is then plotted as error bars in the results. For a specific measurement of $T_{db}$ and $T_{wb}$ at 30 °C and 27 °C, respectively and their respective uncertainties at $\delta T_{db} = \delta T_{wb} = \pm 0.15$°C, $\omega$, $\delta\omega$, and the percentage error can be computed as 21.33 g/kg, ± 0.28 g/kg, and 1.31%, respectively.

$$\delta\omega = \sqrt{\left(\frac{\partial\omega}{\partial T_{db}}\right)^2 (\delta T_{db})^2 + \left(\frac{\partial\omega}{\partial T_{wb}}\right)^2 (\delta T_{wb})^2} \tag{3.13}$$

$$\text{Percentage error} = \pm\frac{\delta\omega}{\omega} \times 100 \tag{3.14}$$

### 3.4.2 Operating Conditions for Performance Testing

The dynamic experimental performance testing of DCHEs is carried out by regulating the ambient conditions or operating parameters. By changing the air temperature and humidity, the performance of DCHEs under different possible outdoor conditions experienced in tropical climates is studied. The operating parameters such as airflow rate, cycle time, cooling water, and hot water temperatures are also carefully controlled to establish the most suitable operating condition of achieving the maximum dehumidification capacity and thermal efficiency. Besides, the performance comparison of coating heat exchangers with conventional and next-generation desiccants is demonstrated. Accordingly, four desiccant materials are used, and Table 3.9 summarizes the experimental facility and different types of parametric studies conducted on them. Tables 3.10, 3.11 and 3.12 specify the list of different

**Table 3.9** Testing facility and performance tests used for studying the performance of DCHEs

| Desiccant coated heat exchangers | Experimental facility | Performance tests |
|---|---|---|
| Silica gel<br>PVA-LiCl (50 w%) | With indoor and outdoor mixing as portrayed in Sect. 3.4.1.2 | • Effect of inlet air temperature<br>• Effect of inlet air humidity ratio |
| SAP-LiCl (50 w%) | With the temperature and humidity chamber as portrayed in Sect. 3.4.1.1 | • Transient performance<br>• Effect of cooling water temperature<br>• Effect of hot water temperature |
| SAP-$HCO_2K$ (50 w%) | With indoor and outdoor mixing as portrayed in Sect. 3.4.1.2 | • Effect of airflow rate<br>• Effect of cycle time |

**Table 3.10** Operating conditions for silica gel and PVA-LiCl (50 w%) performance testing

| Parameters | Units | Baseline conditions | Change range |
|---|---|---|---|
| Cycle time ($t_{cyc}$) | min | 5 (silica gel)<br>10 (PVA-LiCl (50 w%) | – |
| Airflow rate ($\dot{m}_a$) | kg/h | 55 | – |
| Inlet air temperature ($T_{a,in}$) | °C | 30 | 30–36 |
| Inlet air humidity ratio $\left(\omega_{a,in}\right)$ | g/kg | 21.5 (dehumidification)<br>10.5 (regeneration) | 17.5–21.5 |
| Cooling water temperature ($T_{cw,in}$) | °C | 30 | – |
| Hot water temperature ($T_{hw,in}$) | °C | 80 | – |
| Water flow rate ($\dot{m}_w$) | kg/min | 4 | – |

**Table 3.11** Ambient conditions and operating parameters for performance study of SAP-$HCO_2K$ (50 w%) coated heat exchangers

| Parameters | Units | Baseline conditions | Change range |
|---|---|---|---|
| Inlet air temperature ($T_{a,in}$) | °C | 30 | – |
| Inlet air humidity ratio $\left(\omega_{a,in}\right)$ | g/kg | 21.5 (dehumidification)<br>10.5 (regeneration) | – |
| Cycle time ($t_{cyc}$) | min | 10 | 5–30 |
| Airflow rate ($\dot{m}_a$) | kg/h | 55 | 35–65 |
| Cooling water temperature ($T_{cw,in}$) | °C | 30 | – |
| Hot water temperature ($T_{hw,in}$) | °C | 40 | – |
| Water flow rate ($\dot{m}_w$) | kg/min | 4 | – |

operating parameters controlled while conducting the experimental analysis [39, 60, 61]. It is worthy to note that while evaluating the effect of a particular condition, the other parameters are maintained at their baseline conditions.

**Table 3.12** Ambient conditions and operating parameters for SAP-LiCl (50 w%) coated heat exchanger performance testing

| Parameters | Units | Baseline conditions | Change range |
|---|---|---|---|
| Inlet air temperature ($T_{a,in}$) | °C | 30 | – |
| Inlet air humidity ratio $(\omega_{a,in})$ | g/kg | 21.5 | – |
| Cooling water temperature ($T_{cw,in}$) | °C | 25 | 15–30 |
| Hot water temperature ($T_{hw,in}$) | °C | 50 | 40–70 |
| Cycle time ($t_{cyc}$) | min | 10 | – |
| Airflow rate ($\dot{m}_a$) | kg/h | 29–35 | – |
| Cooling/Hot water flow rate ($\dot{m}_w$) | L/min | 4 | – |

### 3.4.3 Performance Indicators

Moisture removal capacity (MRC) is a key performance parameter that evaluates desiccant dehumidifier's capacity. It measures the average amount of moisture removed by the DCHE for a given cycle time and is defined as,

$$\mathrm{MRC} = \frac{1}{t_{cyc}} \int_0^{t_{cyc}} \left(\omega_{a,in} - \omega_{a,out}\right) dt \tag{3.15}$$

where $\omega_{a,in}$ and $\omega_{a,out}$ are the respective air humidity ratio values at inlet and outlet of the DCHE chamber during dehumidification process.

The energy efficiency of DCHEs is computed in terms of thermal coefficient of performance ($\mathrm{COP_{th}}$), which is defined as the ratio of the average cooling capacity of air $\left(\dot{Q}_{a,t}\right)$ to the total heat exchange rate of cold/hot water during dehumidification/regeneration processes $\left(\dot{Q}_w\right)$. $\dot{Q}_{a,t}$ is calculated using Eq. (3.17) and $\dot{Q}_w$ is computed via Eq. (3.21).

$$\mathrm{COP_{th}} = \frac{\dot{Q}_{a,t}}{\dot{Q}_w} \tag{3.16}$$

$$\dot{Q}_{a,t} = \dot{m}_a\left(h_{a,in} - h_{a,out}\right) \tag{3.17}$$

where $h_{a,in}$ and $h_{a,out}$ represent the specific enthalpy of process air at inlet and outlet, respectively, during the dehumidification process. $\dot{Q}_{a,t}$ is also referred as the total cooling load removed from moist air.

The sensible and latent components of the total cooling load are computed via Eqs. (3.18)–(3.20).

$$\dot{Q}_{a,t} = \dot{Q}_{a,s} + \dot{Q}_{a,l} \tag{3.18}$$

$$\dot{Q}_{a,s} = \dot{m}_a c_{p,a} \left( T_{a,in} - T_{a,out} \right) \tag{3.19}$$

$$\dot{Q}_{a,l} = \dot{m}_a h_e \left( \omega_{a,in} - \omega_{a,out} \right) \tag{3.20}$$

where $\dot{Q}_{a,s}$ and $\dot{Q}_{a,l}$ are the respective sensible, latent, and total cooling loads reduced by DCHE; $\dot{m}_a$ is the mass flow rate of air; $c_{p,a}$ is the specific heat capacity of air; $T_{a,in}$ and $T_{a,out}$ are the inlet and outlet air temperatures during dehumidification process, respectively; $h_e$ is the latent heat of vaporization of water evaluated at average dehumidification temperature; and $\omega_{a,in}$ and $\omega_{a,out}$ are the respective inlet and outlet air humidity ratios during the dehumidification process.

$$\dot{Q}_w = \dot{Q}_{cw} + \dot{Q}_{hw} \tag{3.21}$$

where $\dot{Q}_{cw}$ and $\dot{Q}_{hw}$ are the individual components of the respective heat transfer rates in cooling water and hot water and are calculated using Eqs. (3.22) and (3.23).

$$\dot{Q}_{\mathrm{cw}} = \dot{m}_w c_{p,w} \left( T_{cw,out} - T_{cw,in} \right) \tag{3.22}$$

$$\dot{Q}_{\mathrm{hw}} = \dot{m}_w c_{p,w} \left( T_{hw,in} - T_{hw,out} \right) \tag{3.23}$$

where $\dot{m}_w$ is the mass flow rate of water; $c_{p,w}$ is the specific heat capacity of water at constant pressure; $T_{cw,in}$, $T_{cw,out}$, $T_{hw,in}$, and $T_{hw,out}$ are the respective inlet and outlet temperatures of cooling water and hot water flowing inside the tubes of heat exchanger.

### 3.4.4 Transient Performance

The transient performance of DCHEs is demonstrated through the experimental results of dehumidification and regeneration processes of SAP-LiCl (50 w%) coated heat exchangers (SLCHEs). Figure 3.17a shows that the outlet air humidity ratio $\left(\omega_{a,out}\right)$ instantaneously drops from 21.51 g/kg and to reach its minimum at around 13 g/kg. At the same time, Fig. 3.17b and c highlight that air and water temperatures rise by 2–3 °C. The initial rapid drop in $\omega_{a,out}$ is followed by a gradual rise in $\omega_{a,out}$. At the end of $t_{cyc}$, $\omega_{a,out}$ attains 16.52 g/kg. The rapid drop in $\omega_{a,out}$ at the onset of dehumidification is attributed to a higher water vapor diffusivity. Since the desiccant is dry, the moisture is transferred from the air to the desiccant swiftly. However, when this desiccant-air interface layer saturates, the moisture transfer is dominated by the rate at which water can diffuse inside the bulk of the desiccant medium. Since the desiccant's moisture diffusivity is usually around 4–6 orders of

**Fig. 3.17** Dynamic dehumidification and regeneration results of inlet and outlet **a** humidity ratio; **b** air temperature; and **c** water temperature for SAP-LiCl (50 w%) coated heat exchanger under the baseline conditions specified in Table 3.12

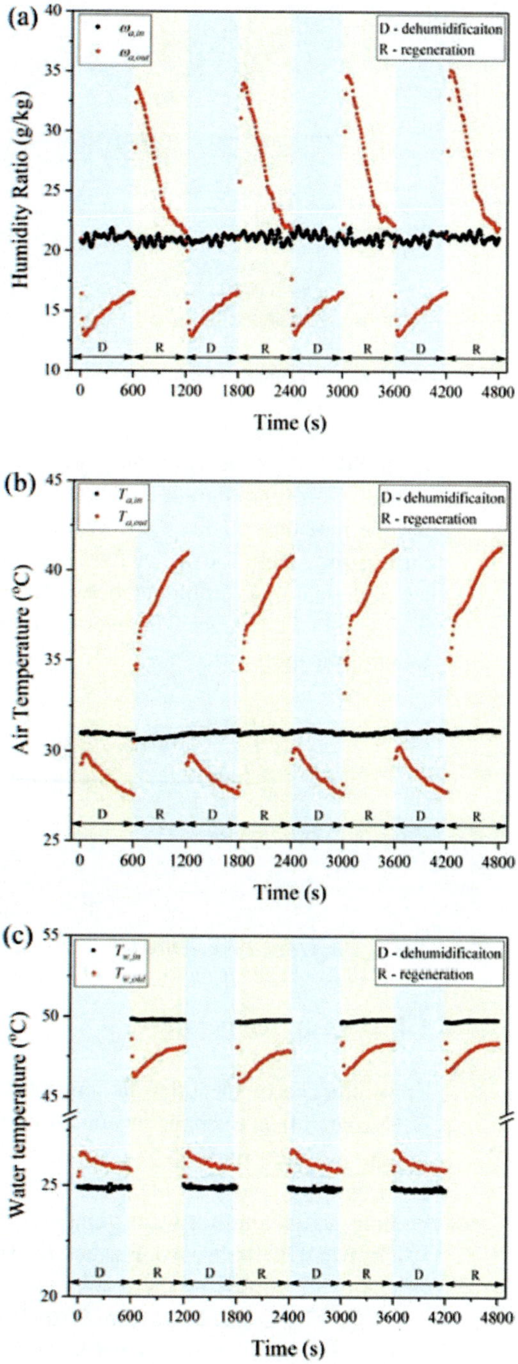

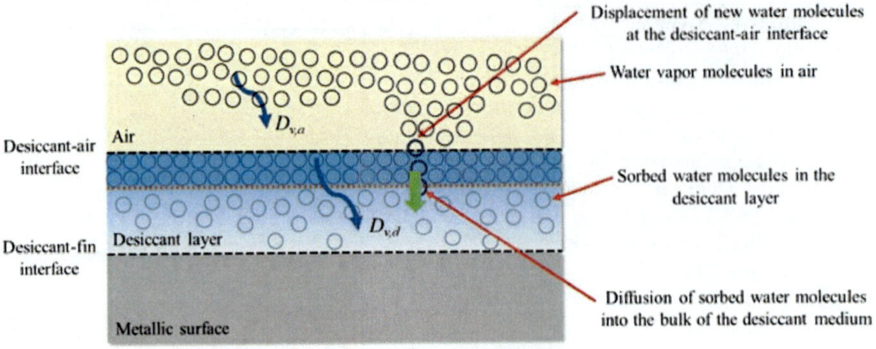


**Fig. 3.18** A schematic representing moisture transfer between air and desiccant layer

magnitude lower than the moisture diffusivity in air (around $10^{-5}$ $m^2/s$) [61], the moisture sorption is gradual. This effect is illustrated in Fig. 3.18. It is noteworthy that the moisture quantity absorbed by the desiccant must be desorbed completely during its regeneration cycle. To verify if appropriate regeneration has occurred, mass balance equation is employed between dehumidification and regeneration processes, as shown in Eq. (3.24). The average mass of water absorbed by the DCHE during dehumidification process is 49.76 g while the amount regenerated is 49.77 g.

$$\underbrace{\int_0^{t_{cyc}} \dot{m}_a\left(\omega_{a,in} - \omega_{a,out}\right)}_{\text{dehumidification}} = \underbrace{\int_0^{t_{cyc}} \dot{m}_a\left(\omega_{a,out} - \omega_{a,in}\right)}_{\text{regeneration}} \tag{3.24}$$

## 3.4.5 *Effect of Climatic Conditions*

### 3.4.5.1 Air Temperature

The influence of the inlet air temperature ($T_{a,in}$) on silica gel and PVA-LiCl (50 w%) coated heat exchangers' dehumidification performance and energy efficiency is demonstrated through Fig. 3.19. The inlet air humidity is maintained constant at 21.51 g/kg, and the other operating parameters such as cycle time, airflow rate, cooling water, and hot water temperatures are also relatively stable at the baseline values listed in Table 3.10. Further, the two DCHEs are abbreviated as SGCHEs and PLCHEs for improving the readability.

When $T_{a,in}$ is regulated from 30 to 36 °C, there is a negative impact of around 10–15% on both SGCHE's and PLCHE's dehumidification capacity. Since the humidity ratio is maintained constant, any rise in air temperature will reduce RH. As the driving for moisture sorption is dependent on RH, a lower RH will affect the desiccant's

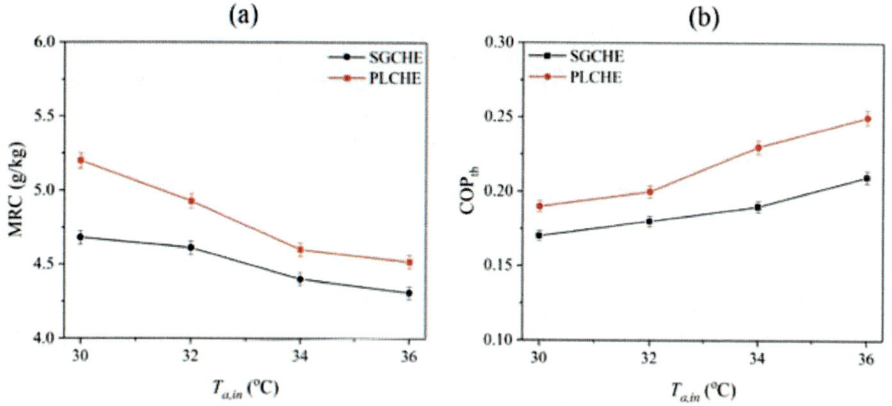


**Fig. 3.19** Effect of $T_{a,in}$ variation from 30 to 36 °C on **a** MRC and **b** $COP_{th}$ of SGCHEs and PLCHEs

dehumidification ability. In contrast, a reduced MRC does not deteriorate the $COP_{th}$. This is because of the higher air enthalpy at elevated air temperatures. Also, a higher gap between $T_{a,in}$ and $T_{cw,in}$ promotes the heat transfer rate and thermal efficiency. While the air temperature plays a key role in the performance ability of DCHEs, the polymer desiccant yields improved dehumidification performance and higher energy efficiency. Its $COP_{th}$ is around 15–20% higher than SGCHEs.

#### 3.4.5.2 Air Humidity Ratio

Figure 3.20 portrays the impact of $\omega_{a,in}$ on the desiccant dehumidifier's MRC and $COP_{th}$. When $\omega_{a,in}$ is reduced from the baseline value of 21.5 g/kg to 17.5 g/kg,

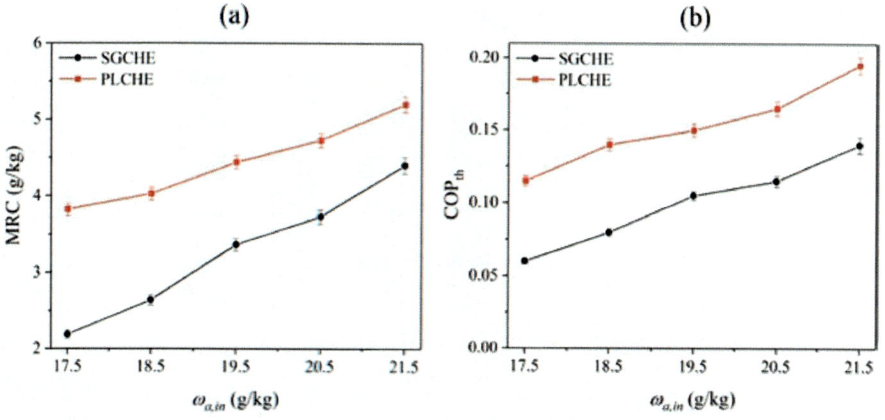


**Fig. 3.20** Effect of inlet air humidity ratio variation from 17.5 to 21.5 g/kg on **a** MRC and **b** $COP_{th}$ of SGCHEs and PLCHEs

MRC drops from 5.2 to 3.8 g/kg for PLCHEs and from 4.4 to 2.19 g/kg for SGCHEs. The reduction in MRC is due to the drop in driving force for moisture sorption at lower humidity content. The reduced dehumidification effect also limits the energy efficiency of the system. $COP_{th}$ of SGCHE and PLCHE decreases by 57% and 40%, respectively. However, the $COP_{th}$ of PLCHEs is around 1–2 times higher than SGCHEs, which indicates the composite desiccant's superior thermal efficiency.

### 3.4.6 Effect of Operating Parameters

#### 3.4.6.1 Airflow Rate

The effect of airflow rate ($\dot{m}_a$) on SAP-$HCO_2K$ coated heat exchanger's (SPFCHE) dehumidification performance and thermal efficiency is shown in Fig. 3.21. It is

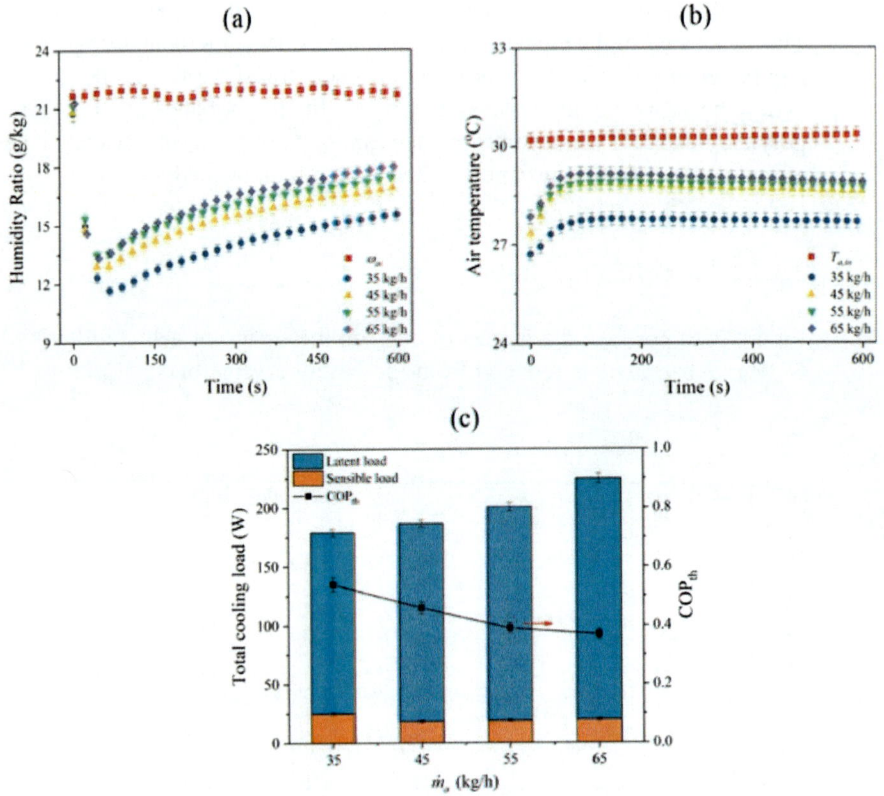


**Fig. 3.21** **a** Dynamic results of inlet and outlet humidity ratio; **b** dynamic results of inlet and outlet air temperature and **c** total cooling load removed and $COP_{th}$ of SPFCHE under the influence of varying airflow rate from 35 to 65 kg/h

readily observed that a lower airflow rate promotes the cooling load removed and enhances its thermal efficiency. This is because when $\dot{m}_a$ is increased, the interaction time between the air and the desiccant ($t_{a,int}$) reduces. This parameter is affected by the DCHE's geometric specification and airflow rate and is computed via Eq. (3.25).

$$t_{a,\text{int}} = \frac{LWH\rho_a}{n\dot{m}_a} \tag{3.25}$$

where $n$ is the number of heat exchangers parallel to the airflow and it is taken as 2 for this study.

The effect of increasing $\dot{m}_a$ from 35 to 65 kg/h reduces $t_{a,\text{int}}$ by two times. While a lower $t_{a,\text{int}}$ is useful to facilitate sufficient time for moisture sorption, the heat exchanger's footprint needs to be sufficiently increased to achieve appropriate mass flow rate. Although Fig. 3.21a points toward improved moisture removal at lower flow rates, Fig. 3.21c shows that the total cooling load, calculated via Eq. (3.18), increases with higher flow rates. The combined effect of around two times rise in $\dot{m}_a$ and about 0.7–0.8 times drop in $\left(\omega_{a,in} - \omega_{a,out}\right)$ translates to higher latent cooling load removal and enhancing the total cooling load. While higher sensible load removal is expected at elevated flow rates, Fig. 3.21b shows that the lowest outlet air temperature is observed at 35 kg/h. The dominating factor affecting this behavior is not apparent. A possible explanation could be the lower quantity of sorption heat production at 35 kg/h because of the reduced absolute mass of water vapor removed.

#### 3.4.6.2 Cycle Time

The effect of a longer cycle time from 5 to 30 min on the performance of SPFCHEs is depicted in Fig. 3.22. When the desiccant is dry at the dehumidification onset, it is

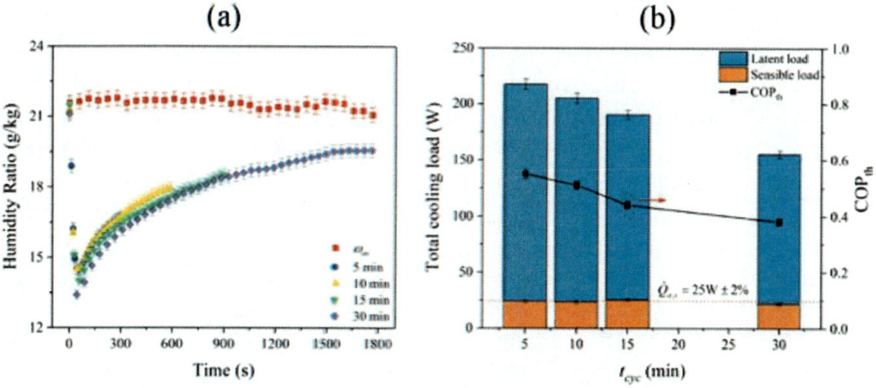


**Fig. 3.22** **a** Dynamic results of inlet and outlet humidity ratio and **b** total cooling load removed and $COP_{th}$ of SPFCHE under the influence of varying cycle time from 5 to 30 min

able to readily capture moisture as the potential for moisture sorption is high. Like the observations made in Sect. 3.4.4, once the desiccant layer interfacing with the air saturates, water molecules need to be transported to the bulk of the medium. Unlike the initial rapid drop in $\omega_{a,out}$, the ensuing process is gradual. As a result, $\omega_{a,out}$ raises steadily, and if the experiment is run for sufficiently long period, it would approach $\omega_{a,in}$. It is worthy to note that every dehumidifier is designed to achieve a desired $\Delta\omega$ for a fixed flow rate. This can only be achieved by appropriately controlling $t_{cyc}$. When $t_{cyc}$ is increased up to 30 min from a short time of 5 min, the average outlet air humidity ratio $(\bar{\omega}_{a,out})$ also rises from around 16–18 g/kg. This results in a significant drop in the latent load removal and depreciates the dehumidifier's energy efficiency by 30%. Although the results favor a shorter cycle time of 5 min, it may not be pragmatic as it results in frequent cycle switching. Higher switching frequency magnifies the effect of thermal mass and would also reduce the lifespan of valves/dampers in the system. As a result, a reasonable trade-off at 10–15 min cycle time is suggested considering the performance and practical aspects.

#### 3.4.6.3 Cooling and Hot Water Temperature

The impact of cooling water and hot water temperatures on dehumidification and thermal performance is examined by conducting experiments on SAP-LiCl (50 w%) coated heat exchangers (SLCHEs). A total of 16 experiments were carried out with cooling water ($T_{cw,in}$) and hot water temperatures ($T_{hw,in}$) regulated between 15 and 30 °C and 40 and 70 °C, respectively. The other operating parameters were maintained at their baseline values as listed in Table 3.12. Figure 3.23a and b show the results of the averaged $\omega_{a,out}$ and $T_{a,out}$ during dehumidification process. The variation in $T_{cw,in}$ markedly affects the $\omega_{a,out}$ and $T_{a,out}$. On an average, the change in $\omega_{a,out}$ and $T_{a,out}$ approach 47.7% and 33.6%, respectively when $T_{cw,in}$ is regulated between 15 and 30 °C. This is because higher temperature gradient between the cooling water and the supply air enhances moisture sorption, and the cooling water captures the sorption heat effectively. In the same vein, the air temperature during the dehumidification process is also reduced at lower $T_{cw,in}$. Further, appropriate regeneration is imperative to achieve effective dehumidification for subsequent cycles. If the regeneration temperature is maintained too low, the heat energy will not be sufficient to overcome the barrier offered by the moisture trapped in the desiccant and will negatively impact the DCHE's dehumidification ability. For SLCHE, the variation of $T_{hw,in}$ from 40 to 70 °C affects $\omega_{a,out}$ and $T_{a,out}$ marginally because SAP-LiCl (50 w%) desiccant is capable of effective regeneration at 40 °C.

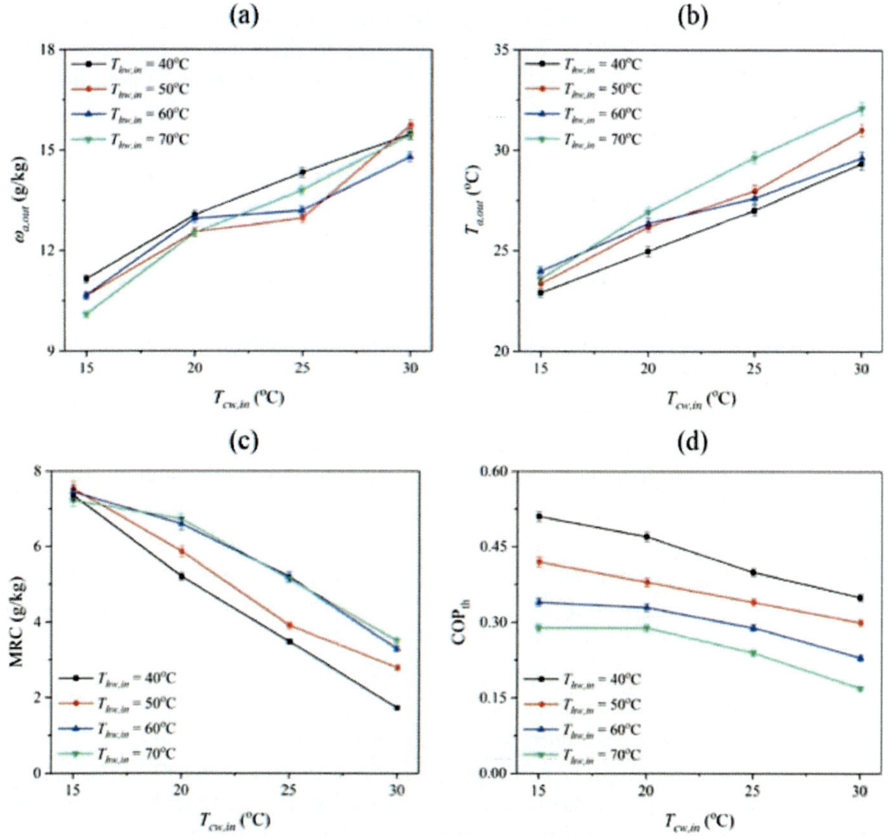


**Fig. 3.23** Effect of variation of $T_{cw,in}$ and $T_{hw,in}$ between 15 and 30 °C and 40 and 70 °C, respectively on **a** outlet air humidity ratio, **b** outlet air temperature, **c** moisture removal capacity (MRC), **d** thermal coefficient of performance ($COP_{th}$)

When $T_{cw,in}$ is higher than the air's dew point temperature, dehumidification occurs only via the desiccant's sorption process. In contrast, reducing $T_{cw,in}$ below the air's dew point facilitates water condensation on the desiccant surface. For an uncoated fin-tube heat exchanger, condensate would begin to form on the metallic surface. However, this does not occur in composite polymer DCHEs since the polymer desiccant absorbs the condensed water. As a result, the moisture removal from the air is due to the confluence of condensation dehumidification and moisture sorption. Accordingly, the highest MRC is observed when $T_{cw,in}$ is maintained 15 °C, with at least 60% improvement compared to 30 °C. Therefore, maintaining a low cooling water temperature is the key to achieve maximum dehumidification performance of SLCHE. It is noteworthy that the $t_{cyc}$ of SLCHE can be maintained over 10 min without any deliquescence even when $T_{cw,in}$ is lower than the air's dew point temperature. Instead, for SGCHEs, the optimum cycle time must be confined to 20–180 s to avoid any undesired deliquescence [62]. Further, Fig. 3.23d highlights that rising

$T_{cw,in}$ and $T_{hw,in}$ impacts the thermal process efficiency negatively. Therefore, maintaining cooling and hot water temperatures to their lowest possible values (while warranting appropriate regeneration) is a pragmatic strategy to maximize $COP_{th}$.

## 3.5 Conclusions

In this chapter, based on the research gaps of obtaining improved desiccant, the synthesis of composite superabsorbent polymer desiccants and MOFs is judiciously discussed. The advantage of these desiccants in terms of yielding superior water uptake capacity, ease of regeneration, and recyclability makes them stand out as next-generation advanced materials compared to conventional desiccants. Although MOFs are able to provide tailorable structural properties and excellent moisture sorption performance, their application is currently limited due to poor cost-effectiveness and difficulties associated with commercialization.

Numerous experimental and theoretical techniques that are available to characterize desiccant materials are listed in this chapter. Experiments are conducted to study the fundamental properties of desiccants and identify the factors that play a critical role in the performance of DCHEs. Accordingly, a higher salt concentration in the composite polymer desiccant is found to enhance its equilibrium water sorption capacity. Nevertheless, the maximum feasible salt concentration is limited to 50 w% to prevent possible occurrence of deliquescence. The composite polymer desiccant is able to offer around 12 times higher isothermal equilibrium capacity compared to conventional silica gel desiccant. In terms of kinetics, around 2–3 times improvement in sorption/desorption rates is obtained. The LDF approximation accurately predicted the sorption and desorption kinetics of composite desiccants. While an increase in salt concentration and temperature improves the sorption and desorption rates, the mechanism behind their increment is found to be different. A higher concentration of the hygroscopic salt significantly promotes water sorption capacity, whereas higher desiccant temperature improves the kinetic constant. Further, the regeneration time is demonstrated not to be a limiting factor in determining the operating time of DCHEs.

Lastly, the influence of operating parameters and ambient conditions on the composite polymer DCHEs' performance is also investigated. The results have revealed that the composite polymer desiccant remains stable even when the cooling water temperature is maintained lower than the dew point temperature of the air. Overall, the composite polymer-based DCHEs is capable of absorbing a substantial amount of moisture at high energy efficiency compared to silica gel-based DCHEs.

## References


1. Tadao S, Takashi N. Preparation and application of high-performance superabsorbent polymers. Superabsorbent Polym. 1994;573:112–27. https://doi.org/10.1021/bk-1994-0573.
2. Czanderna AW, Neidlinger HH. Polymers as advanced materials for desiccant applications:1988 1988.
3. Czanderna AW. Polymers as advanced materials for desiccant applications:1987 1988.
4. Czanderna AW, Tillman NT, Herdt GC. Polymers as advanced materials for desiccant applications part 3: alkali salts of PSSA and polyAMPSA and copolymers of polyAMPSASS. ASHRAE Trans. 1995; 697–712.
5. Czanderna AW. Polymers as advanced materials for desiccant applications: progress report for 1989 1990.
6. White SD, Goldsworthy M, Reece R, Spillmann T, Gorur A, Lee DY. Characterization of desiccant wheels with alternative materials at low regeneration temperatures. Int J Refrig. 2011;34:1786–91. https://doi.org/10.1016/j.ijrefrig.2011.06.012.
7. Higashi T, Zhang L, Saikawa M, Yamaguchi M, Dang C, Hihara E. Études Théoriques Et Expérimentales Sur Les Caractéristiques D'Adsorption Et De Désorption Isothermes D'Un Échangeur De Chaleur Enrobé De Déshydratant. Int J Refrig. 2017;84:228–37. https://doi.org/10.1016/j.ijrefrig.2017.09.007.
8. Lee J, Lee D-Y. Sorption characteristics of a novel polymeric desiccant. Int J Refrig. 2012;35:1940–9. https://doi.org/10.1016/j.ijrefrig.2012.07.009.
9. Yang Y, Rana D, Lan CQ. Development of solid super desiccants based on a polymeric superabsorbent hydrogel composite. RSC Adv. 2015;5:59583–90. https://doi.org/10.1039/C5RA04346H.
10. Bui DT, Nida A, Ng KC, Chua KJ. Water vapor permeation and dehumidification performance of poly(vinyl alcohol)/lithium chloride composite membranes. J Memb Sci. 2016;498:254–62. https://doi.org/10.1016/j.memsci.2015.10.021.
11. Bui TD, Wong Y, Thu K, Oh SJ, Kum Ja M, Ng KC, et al. Effect of hygroscopic materials on water vapor permeation and dehumidification performance of poly(vinyl alcohol) membranes. J Appl Polym Sci. 2017;134:1–9. https://doi.org/10.1002/app.44765.
12. Yao Y, Dai L, Jiang F, Liao W, Dong M, Yang X. Kinetic modeling of novel solid desiccant based on PVA-LiCl electrospun nanofibrous membrane. Polym Test. 2017;64:183–93. https://doi.org/10.1016/j.polymertesting.2017.10.008.
13. Zhang LZ, Wang YY, Wang CL, Xiang H. Synthesis and characterization of a PVA/LiCl blend membrane for air dehumidification. J Memb Sci. 2008;308:198–206. https://doi.org/10.1016/j.memsci.2007.09.056.
14. Rieth AJ, Yang S, Wang EN, Dincă M. Record atmospheric fresh water capture and heat transfer with a material operating at the water uptake reversibility limit. ACS Cent Sci. 2017;3:668–72. https://doi.org/10.1021/acscentsci.7b00186.
15. Abdulhalim RG, Bhatt PM, Belmabkhout Y, Shkurenko A, Adil K, Barbour LJ, et al. A fine-tuned metal-organic framework for autonomous indoor moisture control. J Am Chem Soc. 2017;139:10715–22. https://doi.org/10.1021/jacs.7b04132.
16. Chen Z, Li P, Zhang X, Li P, Wasson MC, Islamoglu T, et al. Reticular access to highly porous acs-MOFs with rigid trigonal prismatic linkers for water sorption. J Am Chem Soc. 2019;141:2900–5. https://doi.org/10.1021/jacs.8b13710.
17. Karmakar A, Prabakaran V, Zhao D, Jon K. A review of metal-organic frameworks (MOFs) as energy-efficient desiccants for adsorption driven heat-transformation applications. Appl Energy. 2020;269: https://doi.org/10.1016/j.apenergy.2020.115070.
18. Yaghi OM, O'Keeffe M, Ockwig NW, Chae HK, Eddaoudi M, Kim J. Reticular synthesis and the design of new materials. Nature. 2003;423:705–14. https://doi.org/10.1038/nature01650.
19. Zu K, Qin M, Cui S. Progress and potential of metal-organic frameworks (MOFs) as novel desiccants for built environment control: A review. Renew Sustain Energy Rev. 2020;133: https://doi.org/10.1016/j.rser.2020.110246.

20. Rubio-Martinez M, Avci-Camur C, Thornton AW, Imaz I, Maspoch D, Hill MR. New synthetic routes towards MOF production at scale. Chem Soc Rev. 2017;46:3453–80. https://doi.org/10.1039/c7cs00109f.
21. Tannert N, Ernst SJ, Jansen C, Bart HJ, Henninger SK, Janiak C. Evaluation of the highly stable metal-organic framework MIL-53(Al)-TDC (TDC = 2,5-thiophenedicarboxylate) as a new and promising adsorbent for heat transformation applications. J Mater Chem A. 2018;6:17706–12. https://doi.org/10.1039/c8ta04407d.
22. Lenzen D, Bendix P, Reinsch H, Fröhlich D, Kummer H, Möllers M, et al. Scalable green synthesis and full-scale test of the metal-organic framework CAU-10-H for use in adsorption-driven chillers. Adv Mater. 2018;30:1705869. https://doi.org/10.1002/adma.201705869.
23. Jeremias F, Khutia A, Henninger SK, Janiak C. MIL-100(Al, Fe) as water adsorbents for heat transformation purposes—a promising application. J Mater Chem. 2012;22:10148–51. https://doi.org/10.1039/C2JM15615F.
24. Demessence A, Horcajada P, Serre C, Boissière C, Grosso D, Sanchez C, et al. Elaboration and properties of hierarchically structured optical thin films of MIL-101(Cr). Chem Commun. 2009:7149–7151. https://doi.org/10.1039/b915011k.
25. Burtch NC, Jasuja H, Walton KS. Water stability and adsorption in metal-organic frameworks. Chem Rev. 2014;114:10575–612. https://doi.org/10.1021/cr5002589.
26. Sohail M, Yun YN, Lee E, Kim SK, Cho K, Kim JN, et al. Synthesis of highly crystalline NH2-MIL-125 (Ti) with s-shaped water isotherms for adsorption heat transformation. Cryst Growth Des. 2017;17:1208–13. https://doi.org/10.1021/acs.cgd.6b01597.
27. Ko N, Choi PG, Hong J, Yeo M, Sung S, Cordova KE, et al. Tailoring the water adsorption properties of MIL-101 metal-organic frameworks by partial functionalization. J Mater Chem A. 2015;3:2057–64. https://doi.org/10.1039/c4ta04907a.
28. Zhang H, Snurr RQ. Computational study of water adsorption in the hydrophobic metal-organic framework ZIF-8: adsorption mechanism and acceleration of the simulations. J Phys Chem C. 2017;121:24000–10. https://doi.org/10.1021/acs.jpcc.7b06405.
29. Xu W, Yaghi OM. Metal-organic frameworks for water harvesting from air, anywhere, anytime. ACS Cent Sci. 2020;6:1348–54. https://doi.org/10.1021/acscentsci.0c00678.
30. Xu F, Bian ZF, Ge TS, Dai YJ, Wang CH, Kawi S. Analysis on solar energy powered cooling system based on desiccant coated heat exchanger using metal-organic framework. Energy. 2019;177:211–21. https://doi.org/10.1016/j.energy.2019.04.090.
31. Zu K, Cui S, Qin M. Performance comparison between metal-organic framework (MOFs) and conventional desiccants (silica gel, zeolite) for a novel high temperature cooling system Recent citations Performance comparison between metal-organic framework (MOFs) and conventional desiccants (silica gel, zeolite) for a novel high temperature cooling system n.d. https://doi.org/10.1088/1757-899X/609/5/052013.
32. Catalyzing Commercialization: Affordable and Sustainable Cooling Using Metal-Organic Frameworksl AIChE n.d. https://www.aiche.org/resources/publications/cep/2020/september/catalyzing-commercialization-affordable-and-sustainable-cooling-using-metal-organic-frameworks. Accessed 11 May 2021.
33. Jeremias F, Fröhlich D, Janiak C, Henninger SK. Water and methanol adsorption on MOFs for cycling heat transformation processes. New J Chem. 2014;38:1846–52. https://doi.org/10.1039/c3nj01556d.
34. Wu H, Salles F, Zajac J. A critical review of solid materials for low-temperature thermochemical storage of solar energy based on solid-vapour adsorption in view of space heating uses. Molecules. 2019;24:945. https://doi.org/10.3390/molecules24050945.
35. Luna-Triguero A, Sławek A, Huinink HP, Vlugt TJH, Poursaeidesfahani A, Vicent-Luna JM, et al. Enhancing the water capacity in Zr-based metal-organic framework for heat pump and atmospheric water generator applications. ACS Appl Nano Mater. 2019;2:3050–9. https://doi.org/10.1021/acsanm.9b00416.
36. Zheng J, Vemuri RS, Estevez L, Koech PK, Varga T, Camaioni DM, et al. Pore-engineered metal-organic frameworks with excellent adsorption of water and fluorocarbon refrigerant for cooling applications. J Am Chem Soc. 2017;139:10601–4. https://doi.org/10.1021/jacs.7b04872.

37. Bae J, Jung JW, Park HY, Cho CH, Park J. Oxygen plasma treatment of HKUST-1 for porosity retention upon exposure to moisture. Chem Commun. 2017;53:12100–3. https://doi.org/10.1039/c7cc05845d.
38. Low JJ, Benin AI, Jakubczak P, Abrahamian JF, Faheem SA, Willis RR. Virtual high throughput screening confirmed experimentally: porous coordination polymer hydration. J Am Chem Soc. 2009;131:15834–42. https://doi.org/10.1021/ja9061344.
39. Vivekh P, Bui DT, Islam MR, Zaw K, Chua KJ. Experimental performance and energy efficiency investigation of composite superabsorbent polymer and potassium formate coated heat exchangers. Appl Energy. 2020;275. https://doi.org/10.1016/j.apenergy.2020.115428.
40. Decoste JB, Peterson GW, Schindler BJ, Killops KL, Browe MA, Mahle JJ. The effect of water adsorption on the structure of the carboxylate containing metal-organic frameworks Cu-BTC, Mg-MOF-74, and UiO-66. J Mater Chem A. 2013;1:11922–32. https://doi.org/10.1039/c3ta12497e.
41. Sindoro M, Jee AY, Granick S. Shape-selected colloidal MOF crystals for aqueous use. Chem Commun. 2013;49:9576–8. https://doi.org/10.1039/c3cc45935g.
42. Bordiga S, Bonino F, Petter Lillerud K, Lamberti C. X-ray absorption spectroscopies: Useful tools to understand metallorganic frameworks structure and reactivity. Chem Soc Rev. 2010;39:4885–927. https://doi.org/10.1039/c0cs00082e.
43. Li A, Thu K, Ismail A Bin, Shahzad MW, Ng KC. Performance of adsorbent-embedded heat exchangers using binder-coating method. Int J Heat Mass Transf. 2016;92:149–57. https://doi.org/10.1016/j.ijheatmasstransfer.2015.08.097.
44. Vivekh P, Bui DT, Wong Y, Kumja M, Chua KJ. Performance evaluation of PVA-LiCl coated heat exchangers for next-generation of energy-efficient dehumidification. Appl Energy. 2019;237. https://doi.org/10.1016/j.apenergy.2019.01.018.
45. C-Therm Thermal Conductivity Instruments|C-Therm Technologies Ltd. n.d. https://ctherm.com/. Accessed 25 Apr 2021.
46. Greathouse JA, Allendorf MD. The interaction of water with MOF-5 simulated by molecular dynamics. J Am Chem Soc. 2006;128:10678–9. https://doi.org/10.1021/ja063506b.
47. Sodium acrylate|C3H3NaO2-PubChem n.d. https://pubchem.ncbi.nlm.nih.gov/compound/Sodium-acrylate. Accessed 23 Oct 2019.
48. Lithium chloride|LiCl-PubChem n.d. https://pubchem.ncbi.nlm.nih.gov/compound/Lithium-chloride. Accessed 23 Oct 2019.
49. Potassium formate|Sigma-Aldrich n.d. https://www.sigmaaldrich.com/catalog/substance/potassiumformate841259029411?lang=en&region=SG. Accessed 7 Feb 2020.
50. Polyvinyl alcohol|CH2CHOH-PubChem n.d. https://pubchem.ncbi.nlm.nih.gov/compound/Polyvinyl-alcohol. Accessed 5 Mar 2020.
51. Burdick CL, Latta JL. Method of using poly(vinyl alcohol) fluidized polymer suspensions in aqueous systems. US5541241A, 1996.
52. Sultan M, El-Sharkawy II, Miyazaki T, Saha BB, Koyama S, Maruyama T, et al. Water vapor sorption kinetics of polymer based sorbents: theory and experiments. Appl Therm Eng. 2016;106:192–202. https://doi.org/10.1016/j.applthermaleng.2016.05.192.
53. Jones B, Ghali K, Ghaddar N. Heat-moisture interactions and phase change in fibrous material. Thermal and Moisture Transport in Fibrous Materials, Elsevier Ltd.;2006. p. 424–436. https://doi.org/10.1533/9781845692261.2.424.
54. Pérez-Alonso C, Beristain CI, Lobato-Calleros C, Rodríguez-Huezo ME, Vernon-Carter EJ. Thermodynamic analysis of the sorption isotherms of pure and blended carbohydrate polymers. J Food Eng. 2006;77:753–60. https://doi.org/10.1016/J.JFOODENG.2005.08.002.
55. Entezari A, Ge TS, Wang RZ. Water adsorption on the coated aluminum sheets by composite materials (LiCl + LiBr)/silica gel. Energy. 2018;160:64–71. https://doi.org/10.1016/j.energy.2018.06.210.
56. Yao Y, Zhang W, Peng Y, Wang L, Liu Y, Chen B. Modeling of silica gel dehydration assisted by power ultrasoni . Refrig Air Cond. 2010:1–9. (* Corresponding Author : Ye Yao).
57. Coleman HW, Steele WG. Uncertainty analysis for uncertainty analysis for engineers 2009.
58. Buck AL. New equation for computing vapor pressure. J Appl Meteorol. 1981.

59. Wood LA. The use of dew-point temperature in humidity calculations. J Res Natl Bur Stand Sect C Eng Instrum. 1970;74C:117. https://doi.org/10.6028/jres.074c.014.
60. Vivekh P, Islam MR, Chua KJ. Experimental performance evaluation of a composite super-absorbent polymer coated heat exchanger based air dehumidification system. Appl Energy 2020;260. https://doi.org/10.1016/j.apenergy.2019.114256.
61. Vivekh P, Bui DT, Wong Y, Kumja M, Chua KJ. Performance evaluation of PVA-LiCl coated heat exchangers for next-generation of energy-efficient dehumidification. Appl Energy. 2019;237:733–50. https://doi.org/10.1016/j.apenergy.2019.01.018.
62. Ge TS, Dai YJ, Wang RZ. Analysis on integrated low grade condensation heat powered desiccant coated vapor compression system. Appl Therm Eng. 2018;138:307–18. https://doi.org/10.1016/j.applthermaleng.2018.04.044.

# Chapter 4
# Advanced Engineering Analysis of Desiccant Coated Dehumidifiers

**Abstract** In contrast to experimental testing, which is time-consuming and expensive, theoretical studies provide an economical and judicious methodology to evaluate the performance of desiccant dehumidifiers. Such approaches, developed based on the fundamental laws of thermodynamics, flow physics, and heat transfer, are critical to improving the dehumidifier design. This chapter firstly presents a general mathematical approach to predict the performance of DCHEs based on the governing principles of mass, momentum, energy, and species conservation. Next, the model is validated with the experimental results for different types of desiccants. Further, the validated model is employed to predict the performance of new fin-tube configurations. Detailed energy and economic analysis are then conducted on a hybrid central air-conditioning system comprising a composite superabsorbent polymer and mechanical vapor compression chillers. The hybrid configuration's electrical power savings and its payback period are computed via this analysis. Lastly, the second law of thermodynamics is applied to study DCHEs and identify the causes of key irreversibility.

**Keywords** CFD · Mathematical model · Energy analysis · Economic analysis · Exergy analysis · Hybrid air-conditioning system

 
V. Prabakaran and K. J. Chua, *Advances in Desiccant Dehumidification*,
https://doi.org/10.1007/978-3-030-80843-3_4

## List of Symbols

| | |
|---|---|
| $A_{sc}$ | Solar collector area, $m^2$ |
| $c$ | Concentration, $mol/m^3$ |
| $c_p$ | Specific heat capacity at constant pressure, $kJkg^{-1}\ K^{-1}$ |
| $D$ | Mass diffusivity, $m^2/s$ |
| $\dot{E}x_{dest}$ | Exergy destruction rate, kW |
| $g$ | Number of heat exchangers in the hybrid system |
| $h$ | Specific enthalpy, kJ/kg |
| $I$ | Solar irradiation, $W/m^2$ |
| $k$ | Kinetic constant, $s^{-1}$/thermal conductivity, $Wm^{-1}\ K^{-1}$ |
| $k_T$ | Turbulent kinetic energy, $m^2/s^2$ |
| $\dot{m}$ | Mass flow rate, kg/s or kg/h |
| $M$ | Molecular weight, g/mol |
| $N_f$ | Number of fins |
| $N_t$ | Number of tube pass |
| $P$ | Pressure, kPa |
| $P_0$ | Atmospheric pressure, kPa |
| $P_{el}$ | Electrical power, kW |
| $P_k$ | Production term in the turbulence model |
| $q$ | Water uptake by the desiccant, kg/kg or % |
| $q_{st}$ | Heat of sorption, kJ/kg |
| $\dot{Q}$ | Heating power absorbed or required, kW |
| $\dot{Q}_{lost}$ | Heat loss rate, kW |
| $s$ | Specific entropy, $kJkg^{-1}\ K^{-1}$ |
| $s_g$ | Specific entropy generation, $Jg^{-1}\ K^{-1}$ |
| $\dot{S}_{gen}$ | Overall entropy generation rate, W/K |
| $t$ | Time, s |
| $T$ | Temperature, °C or K |
| $t_{cyc}$ | Cycle time, min |
| $\boldsymbol{u}$ | Velocity vector, m/s |
| $V$ | Velocity, m/s |
| $\dot{W}_{actual}$ | Actual work input in the DCHE process, kW |
| $\dot{W}_{rev}$ | Reversible work input for an ideal air-conditioning process, kW |
| *Greek Symbols* | |
| $\varepsilon_T$ | Turbulent dissipation rate, $Jkg^{-1}\ s^{-1}$ |
| $\xi$ | Specific flow exergy, kJ/kg |
| $\Delta H_s$ | Heat of sorption, kJ/kg |
| $\eta$ | Efficiency |

(continued)

| | |
|---|---|
| $\eta_{II}$ | Second law efficiency |
| $\rho$ | Density, g/cm$^3$ or kg/m$^3$ |
| $\varphi$ | Relative humidity, % |
| $\Phi$ | Flux |
| $\tau$ | Number of hours of operation, hr |
| $\mu$ | Dynamic viscosity, Pa.s |
| $\mu_T$ | Turbulent viscosity, Pa.s |
| $v$ | Specific volume, m$^3$/kg |
| $\omega$ | Humidity ratio, g/kg |
| $\tilde{\omega}$ | Mole fraction ratio of vapor to air, mol/kmol |
| *Subscripts* | |
| 0 | Initial time/dead state in exergy analysis |
| a | Air |
| amb | Ambient |
| atm | Atmosphere |
| cc | Cooling coil |
| ch | Chiller |
| CT | Cooling tower |
| cw | Cooling water |
| cyc | Cycle |
| d | Desiccant/dehumidification |
| e | Equilibrium |
| el | Electrical |
| f | Saturated liquid state |
| g | Saturated vapor state |
| hw | Hot water |
| in | Inlet |
| max | Maximum |
| min | Minimum |
| out | Outlet |
| r | Regeneration |
| rev | Reversible |
| sat | Saturated |
| sim | Simulation |
| sc | Solar collector |
| v | Vapor |
| w | Liquid water |

(continued)

*Abbreviations*

| | |
|---|---|
| AHU | Air handling unit |
| Al | Aluminum |
| ASHRAE | American Society of Heating, Refrigerating, and Air-conditioning Engineers |
| CAPEX | Capital expenditure, US$ |
| CCF | Cumulative cash flow, US$ |
| CFCs | Chlorofluorocarbons |
| $CO_2e$ | Carbon dioxide-equivalent |
| COE | Cost of electricity |
| COP | Coefficient of performance |
| $COP_{th}$ | Thermal coefficient of performance |
| Cu | Copper |
| DCHE | Desiccant coated heat exchanger |
| EPS | Electrical power saved, % |
| IRR | Internal rate of return |
| NCF | Net cash flow, US$ |
| NRMSE | Normalized root mean square error |
| OPEX | Operational expenditure, US$ |
| PLCHE | PVA-LiCl (50 w%) coated heat exchanger |
| RH | Relative humidity |
| SAP | Super absorbent polymer |
| SGCHE | Silica gel coated heat exchanger |
| SLCHE | SAP-LiCl (50 w%) coated heat exchanger |
| SPFCHE | SAP-$HCO_2K$ (50 w%) coated heat exchanger |

## 4.1 Introduction

In the previous chapter, developments in advanced desiccant synthesis and experimental performance studies have been conducted to obtain dehumidification and thermal performance of DCHE dehumidifiers. As costs incurred from tedious experimentations are higher and achieving steady-state operating conditions is time-consuming, theoretical analyses are essential to evaluate new avenues for improving DCHE's design. In this chapter, a computational fluid dynamics (CFD) approach is introduced to predict the dynamic performance of DCHEs and compute the temperature/humidity distributions. Next, a hybrid air-conditioning system with composite superabsorbent polymer DCHE is considered for evaluating the energy benefits. The first law of thermodynamics coupled with economic analysis is carried out to estimate the hybrid air-conditioning system's energy and cost savings potential. Lastly, a judiciously crafted exergy balance analysis is derived from the experimental findings to improve the understanding of various irreversibility associated with DCHEs.

## 4.2 Theoretical Performance Approach

A mathematical methodology is first proposed to simulate the transient performance of DCHEs based on the fundamental laws of mass, momentum, energy, and species conservation principles [1]. The model,

i. takes into account of fluid flow, temperature, and moisture distributions across different domains,
ii. incorporates a three-dimensional system with the cross flow of air and water,
iii. models the moisture diffusion in the desiccant bulk layer and includes realistic conductive and convective heat transfer resistances.

Ideally, a complete DCHE needs to be chosen as the domain in order to achieve a high level of accurate simulated results. Since such an evaluation leads to solving the governing equations employing millions of finite elements, it imposes pronounced computational difficulties. To simplify the analysis, the simulating domain is reduced to a single fin-tube path. This assumption is valid since there are several axes of symmetry for energy and species transport, as portrayed in Fig. 4.1. The reduced model consists of numerous domains such as the Cu tube, Al fin, airflow, water

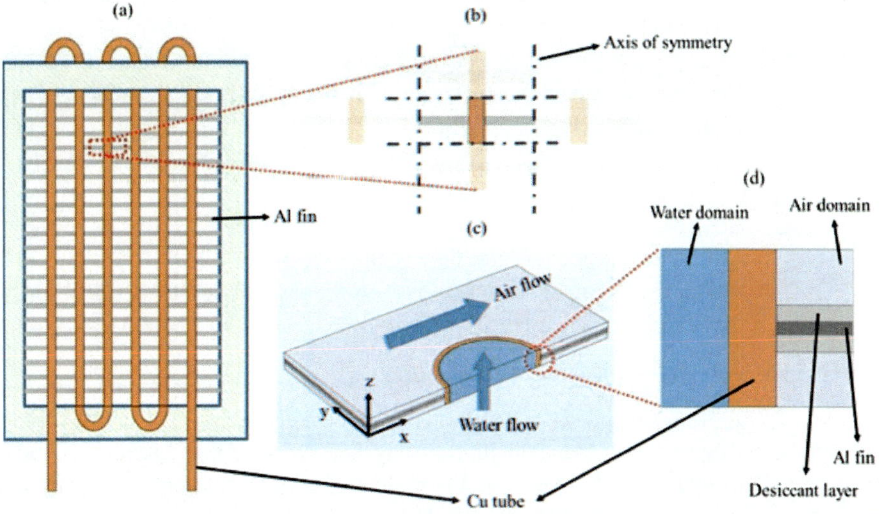


**Fig. 4.1** Schematic of the **a** actual uncoated fin-tube heat exchanger; **b** uncoated single-fin tube approximation highlighting the symmetries involved in the fin-tube heat exchanger; **c** three-dimensional representation of the single fin-tube coated heat exchanger; and **d** a magnified version of the DCHE highlighting different domains

flow, and the desiccant coated layer on the fins. The airflow and the water flow are respectively modeled along X-axis and Z-axis to represent the crossflow direction. A single longitudinal tube-pass heat exchanger is chosen for experimental validation so that the airflow path for the reduced model remains identical to the actual heat exchanger. However, due to the domain reduction, the water flow path is modeled only for a small portion. To derive the water temperature of the heat exchanger from the model, the temperature rise across each fin-tube path is assumed to be linear. Again, this assumption is suitable because the DCHE's operating time is around 2–10 min, and the highest limit for water flow temperature increase/decrease ranges spanning 3–5 °C [2]. The water temperatures are then computed via

$$T_{cw,in} - T_{cw,out} = \left(T_{cw,in} - T_{cw,out}\right)_{sim} N_f N_t \tag{4.1}$$

$$T_{hw,in} - T_{hw,out} = \left(T_{hw,in} - T_{hw,out}\right)_{sim} N_f N_t \tag{4.2}$$

### 4.2.1 Assumptions

The following assumptions are employed to simplify the mathematical model:

1. A uniform desiccant coating layer is considered for the fins and desiccant coating over the tubes is neglected. This is because the effective fin area available for coating ($0.7\text{m}^2$) is substantially larger than the corresponding tube area ($0.035\text{m}^2$).
2. Al fins, Cu tube, and desiccant layer are isotropic materials, and their thermo-physical properties: density, specific heat capacity, and thermal conductivity are assumed to behave as constants.
3. The release/absorption of sorption/desorption heat occurs at the interface between the air and the desiccant layer.
4. Heat losses from the fin tips/ends and the radiation effects are deemed negligible.
5. A fully developed laminar flow regime represents the airflow domain [3], and a Reynolds-averaged Navier–Stokes (RANS) $k-\varepsilon$ turbulence model is employed for modeling the water flow in the tubes. The entrance length effects are assumed to be insignificant.
6. The effect of contact resistance between fins and tubes is considered negligible as compared to other heat transfer resistances [4].
7. Table 4.1 represents the thermo-physical properties of different DCHE domains. In this analysis, the model is validated with the experimental observations of silica gel and PVA-LiCl (50 w%) coated heat exchangers, which are abbreviated as SGCHEs and PLCHEs for simple referencing. The thermophysical properties are assumed to be constants for silica gel [5, 6], PVA-LiCl [7], Al, and Cu

**Table 4.1** Thermophysical properties of different domains in the DCHE system

| Property | Units | Silica gel | PVA-LiCl | Air | Water | Al | Cu |
|---|---|---|---|---|---|---|---|
| Thermal conductivity ($k$) | $Wm^{-1} K^{-1}$ | 0.198 | 0.31 | $k_a(T)$ | $k_w(T)$ | 238 | 400 |
| Density ($\rho$) | $kg/m^3$ | 700–800 | 1521 | $\rho_a(T)$ | $\rho_w(T)$ | 2700 | 8960 |
| Specific heat capacity at constant pressure ($c_p$) | $kJkg^{-1} K^{-1}$ | 0.921 | 1.4 | $c_{p,a}(T)$ | $c_{p,w}(T)$ | 0.9 | 0.385 |
| Dynamic viscosity ($\mu$) | Pa.s | – | – | $\mu_a(T)$ | $\mu_w(T)$ | – | – |

in the operating range. Having said that, for air and water, the variation of thermophysical properties with respect to temperature is adopted from the standard handbooks [8].

### 4.2.2 Governing Equations

Equations (4.3)–(4.6) represent the continuity, momentum, energy, and species conservation equations in the air domain.

$$\rho_a(\nabla \boldsymbol{u}_a) = 0 \tag{4.3}$$

$$\rho_a\left[\frac{\partial \boldsymbol{u}_a}{\partial t} + (\boldsymbol{u}_a \cdot \nabla)\boldsymbol{u}_a\right] = -\nabla P_a + \mu_a \Delta \boldsymbol{u}_a \tag{4.4}$$

$$\rho_a c_{p,a}\left(\frac{\partial T_a}{\partial t} + \boldsymbol{u}_a \cdot \nabla T_a\right) + \nabla \cdot (-k_a \nabla T_a) = 0 \tag{4.5}$$

$$\frac{\partial c_a}{\partial t} + \nabla \cdot (-D_a \nabla c_a) + \boldsymbol{u}_a \cdot \nabla c_a = 0 \tag{4.6}$$

For the desiccant, Eqs. (4.7) and (4.8) represent the energy and species conservation phenomena. Since flow or exchange of species does not occur in aluminum fins and copper tube, the energy conservation equation alone is used as shown in Eqs. (4.9) and (4.10). The continuity and momentum conservation equations for water flowing in the tubes is denoted using RANS $k - \varepsilon$ turbulence model, as shown in Eqs. (4.11)–(4.14) and its energy transfer is given in Eq. (4.15).

$$\rho_d c_{p,d} \frac{\partial T_d}{\partial t} + \nabla \cdot (-k_d \nabla T_d) = 0 \tag{4.7}$$

$$\frac{\partial c_d}{\partial t} + \nabla \cdot (-D_d \nabla c_d) = 0 \tag{4.8}$$

$$\rho_{Al} c_{p,Al} \frac{\partial T_{Al}}{\partial t} + \nabla \cdot (-k_{Al} \nabla T_{Al}) = 0 \tag{4.9}$$

$$\rho_{Cu} c_{p,Cu} \frac{\partial T_{Cu}}{\partial t} + \nabla \cdot (-k_{Cu} \nabla T_{Cu}) = 0 \tag{4.10}$$

$$\rho_w (\nabla \boldsymbol{u}_w) = 0 \tag{4.11}$$

$$\rho_w \left[ \frac{\partial \boldsymbol{u}_w}{\partial t} + (\boldsymbol{u}_w \cdot \nabla) \boldsymbol{u}_w \right] = -\nabla P_w + \left( \mu_w + \mu_{T,w} \right) \Delta \boldsymbol{u}_w \tag{4.12}$$

$$\rho_w \left( \frac{\partial k_{T,w}}{\partial t} + (u_w \cdot \nabla) k_{T,w} \right) = \nabla \cdot \left[ \left( \mu_w + \mu_{T,w} \right) \nabla k_{T,w} \right] + P_{k,w} - \rho_w \varepsilon_{T,w} \tag{4.13}$$

$$\rho_w \left( \frac{\partial \varepsilon_{T,w}}{\partial t} + (u_w \cdot \nabla) \varepsilon_{T,w} \right) = \nabla \cdot \left[ \left( \mu_w + \frac{\mu_{T,w}}{1.3} \right) \nabla \varepsilon_{T,w} \right] + 1.44 \frac{\varepsilon_{T,w}}{k_{T,w}} P_{k,w} - 1.92 \rho_w \frac{\varepsilon_{T,w}^2}{k_{T,w}} \tag{4.14}$$

$$\rho_w c_{p,w} \left( \frac{\partial T_w}{\partial t} + \boldsymbol{u}_w \cdot \nabla T_w \right) + \nabla \cdot (-k_w \nabla T_w) = 0 \tag{4.15}$$

As discussed earlier in Chap. 3, the equilibrium sorption capacity ($q_e$) is a desiccant characteristic required to support the governing equations. $q_e$ is the maximum amount of moisture that a desiccant can adsorb/absorb and it is only a function of relative humidity ($\varphi$) [9], as represented by Eq. (4.16). The characteristic isotherm equation for a desiccant is obtained by regressing the gravimetric experimental data with a second or third-order polynomial equation [10]. The maximum moisture concentration of the desiccant is deduced by.

$$q_e = f(\varphi) \tag{4.16}$$

$$c_{d,\max} = \frac{\rho_d q_e}{100 M_w} \tag{4.17}$$

where $q_e$ is the equilibrium amount of the moisture absorbed by the desiccant, and $\varphi$ is the RH.

Further, the LDF approximation that represents desiccant kinetics [11] is shown in Eq. (4.18) and the relations for mass and heat fluxes at the desiccant-air interface are presented in Eqs. (4.19) and (4.20).

$$\frac{dq_t}{dt} = k(q_e - q_t) \tag{4.18}$$

$$\phi_m = k\delta_d\left(c_{d,\max} - c_d\right) \tag{4.19}$$

$$\phi_h = \phi_m \Delta H \tag{4.20}$$

where $q_t$ is the instantaneous mass absorbed/adsorbed by desiccant, $q_e$ is the equilibrium mass absorbed/adsorbed by desiccant, $q_0$ is the initial water content present in desiccant, and $k$ is the kinetic constant.

The diffusivity of water vapor in air is computed by Hall and Pruppacher correlation [12].

$$D_a = 2.11 \times 10^{-5}\left(\frac{T}{273.15}\right)^{1.94}\left(\frac{P_0}{P}\right) \tag{4.21}$$

### 4.2.3 Numerical Methodology and Mesh Independence Test

The mathematical model is combined with appropriate initial and boundary conditions [10], and is simulated using the COMSOL MultiPhysics v5.3 software platform [13]. This platform uses finite element method to discretize the governing equations. The mesh scheme of the model geometry is shown in Fig. 4.2a. A mesh independence test is carried out by implementing 4 mesh schemes with increasing mesh precision. Accordingly, mesh 1 is coarser with 63,984 elements and mesh 4 is the finest quality with 3,102,625 elements. Figure 4.2b and c demonstrate that the outlet air temperature and humidity predicted by the model do not vary significantly with respect to mesh quality. Therefore, mesh 1 is considered appropriate and adopted for all the analysis.

### 4.2.4 Validation of the Mathematical Model

The mathematical model is next validated with the experimental data acquired from the experimental setup under dehumidification and regeneration cycles of SGCHEs and PLCHEs. Table 4.2 specifies the list of experimental operating parameters used for validation.

In Fig. 4.3a, the $\omega_{a,out}$ of SGCHE drops steeply from 21.5 to 9.8 g/kg in 30 s and then gradually moves towards the inlet 21.5 g/kg at 310 s. At 600 s, the dehumidification mode of SGCHE is shifted to regeneration, where the $\omega_{a,out}$ first rises to 26 g/kg and then progressively declines to match with its inlet value. Figure 4.3b illustrates that the outlet air temperature $\left(T_{a,out}\right)$ also rises from 30 to 33.2 °C at 30 s, and then gradually depreciates to 31.8 °C at 600 s. Likewise, during the regeneration process, as the hot water is maintained at 80 °C, the air temperature sharply increases

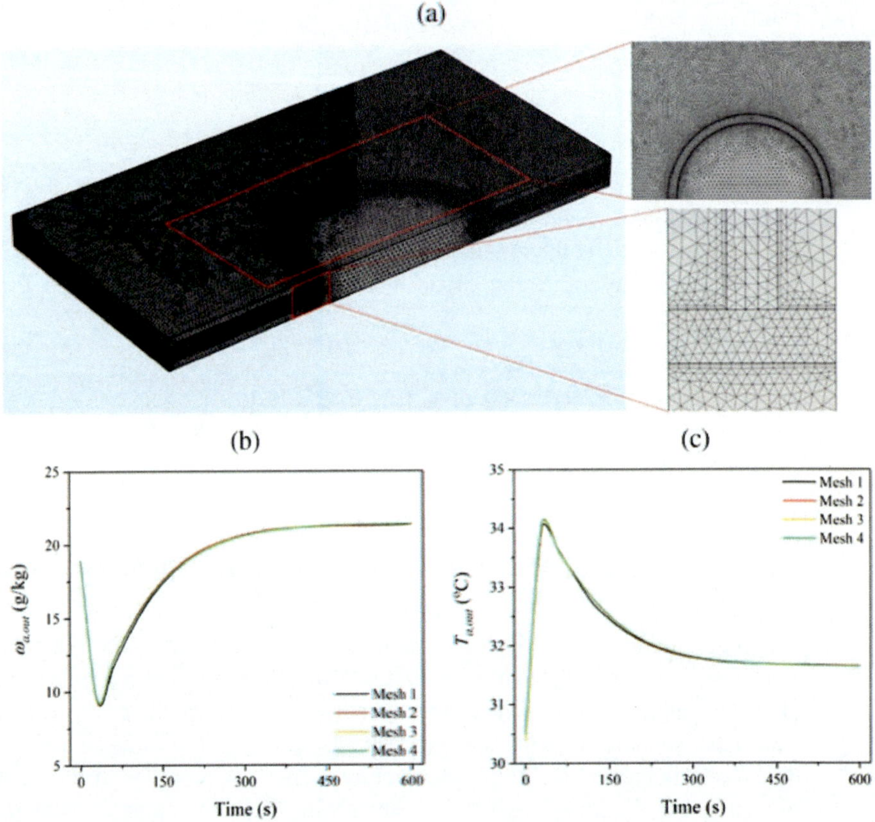


**Fig. 4.2** **a** Mesh scheme of the DCHE model; the results of mesh independence test using four different mesh schemes for **b** outlet air humidity ratio and **c** outlet air temperature

**Table 4.2** Operating conditions for DCHE performance testing

| Parameters | Units | Experimental conditions |
|---|---|---|
| Cooling water temperature ($T_{cw,in}$) | °C | 30 |
| Hot water temperature ($T_{hw,in}$) | °C | 80 |
| Cycle time ($t_{cyc}$) | min | 10 |
| Airflow rate ($\dot{m}_a$) | kg/h | 55 |
| Inlet air temperature ($T_{a,in}$) | °C | 30 |
| Inlet air humidity ratio ($\omega_{a,in}$) | g/kg | 21.5 |
| Water flow rate | L/min | 4 |

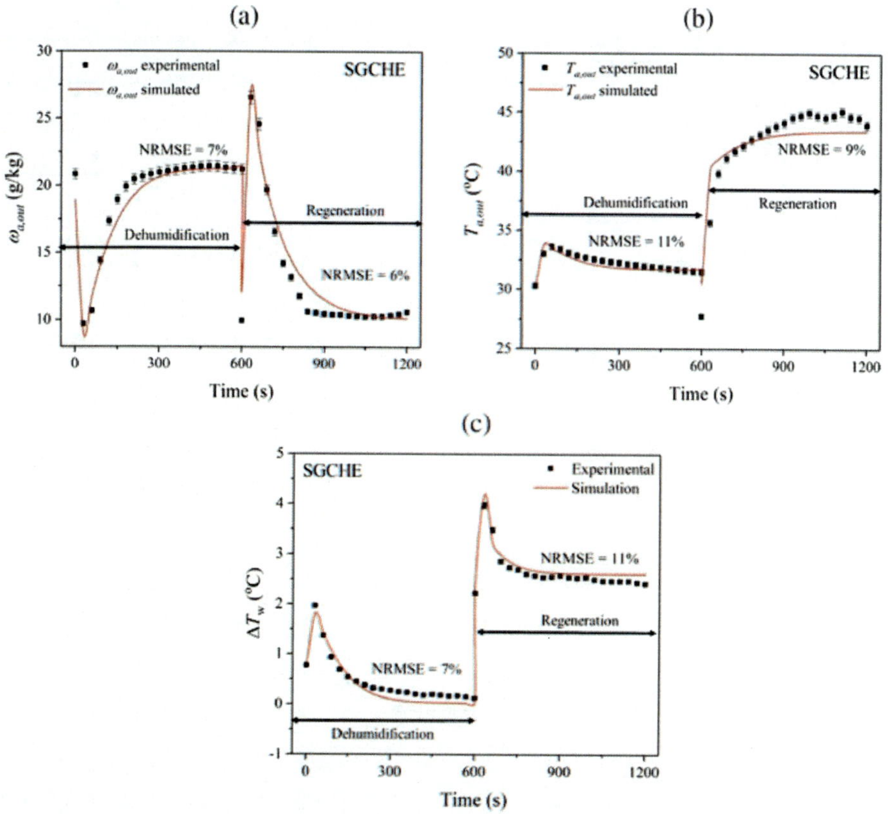


**Fig. 4.3** Experimental results and model validation of the transient response of **a** outlet air humidity ratio, **b** outlet air temperature, and **c** difference in the water temperature of SGCHE under the baseline conditions of dehumidification and regeneration modes

to 40 °C and then followed by gradual rise until 44.5 °C. The change in the water temperature ($\Delta T_w$), captured by the mathematical model and experiment, is shown in Fig. 4.3c. Since the model captures the water temperature across only a single fin, Eqs. (4.1) and (4.2) are used to evaluate $\Delta T_w$ for the actual DCHE. As demonstrated in Fig. 4.3c, the model accurately depicts the trend of cooling water temperature with the highest discrepancy of ±0.6 °C. The dynamic behavior of $\omega_{a,out}$, $T_{a,out}$, and $\Delta T_w$ for PLCHE is shown in Fig. 4.4, and the behavior is similar to SGCHEs. The simulation results agree well with experimental data for both SGCHE and PLCHE, albeit small discrepancies occurring towards the end of the experiment.

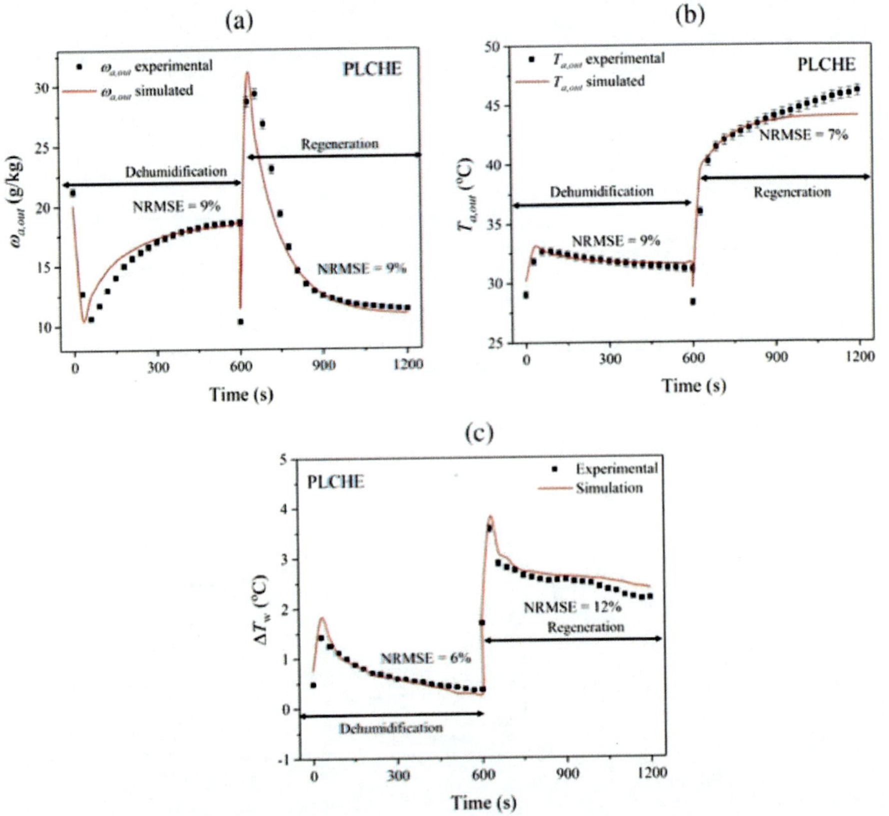


**Fig. 4.4** Experimental results and model validation of the transient response of **a** outlet air humidity ratio, **b** outlet air temperature, and **c** difference in the water temperature of PLCHE under the baseline conditions of dehumidification and regeneration modes

### *4.2.5 Parametric Study on the Effect of Fin-Tube Configurations*

The mathematical model can be utilized to study the flow field distributions and to determine the localized functions of desiccant's temperature and moisture concentration. Further, the effect of geometric parameters and desiccant's moisture sorption characteristics can be obtained by employing the mathematical model. In this section, the impact of different fin-tube configurations on DCHEs' performance is analyzed. Silica gel DCHE with two rectangular fins, an annular fin, and no fin configurations are considered. Their schematics and salient features are provided in Fig. 4.5 and Table 4.3, respectively.

Figure 4.6 illustrates the transient response of $\omega_{a,out}$ and $T_{a,out}$ for different DCHE designs. The peak $\omega_{a,out}$ values for two and three fin-tube tube configurations demonstrate at least 15% improvement as compared to the existing single fin-tube type.

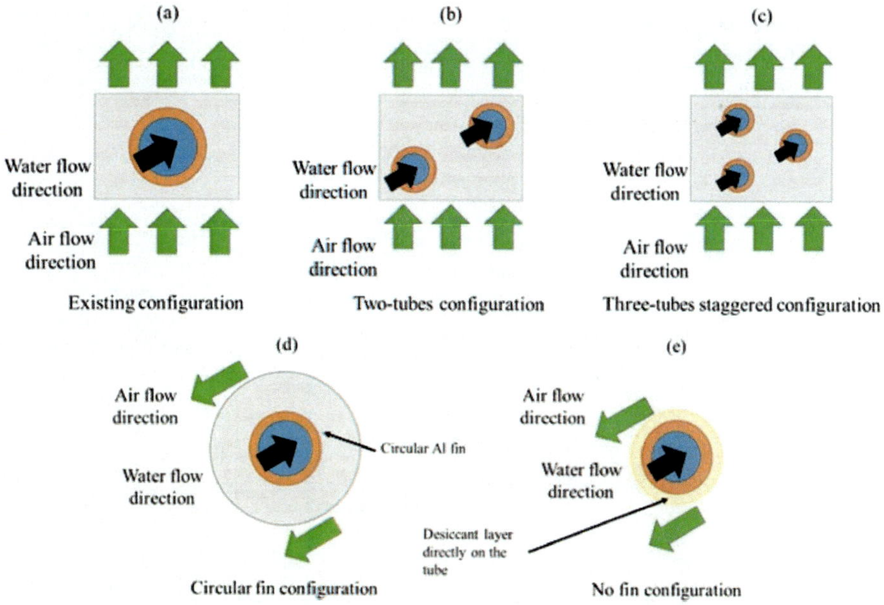


**Fig. 4.5** Schematic of **a** existing single fin-tube configuration; **b** two-tubes configuration; **c** three-tube staggered configuration; **d** annular fin configuration; and **e** no-fin configuration, where the desiccant is directly coated on the tube

Contrastingly, the peak $\omega_{a,out}$ of the annular fin-tube configuration deteriorates by 35%. While the single-fin configuration undergoes a rapid appreciation, the rate of increase of $\omega_{a,out}$ for the annular fin configuration is 45% lower than the existing configuration. As a result, the annular configuration produces improved dehumidification effect. Lastly, the peak $\omega_{a,out}$ for the no fin design is identical to the single fin-tube configuration, and the appreciation thereafter is more gradual unlike the single-fin configuration. Further, due to improved heat transfer effect, the peak $T_{a,out}$ of all the new configurations show around 10% reduction.

Figure 4.7 compares the MRC of the new and existing configurations. The no-fin configuration records 20% higher MRC compared to the single fin-tube configuration. Yet, due to its significantly lower coating area, the compactness factor of the dehumidifier is negatively affected. The two-tube configuration records a 7% higher MRC due to the improved heat transfer from the additional tube source. This second tube improves the heat transfer distribution and captures the sorption heat more effectively. Also, it is apparent that MRC improves only marginally by <1% by adding a third tube, which implies that the two-tube configuration is more suitable. Lastly, the MRC of the annular fin configurations is around 40% higher than the existing configuration of DCHE. Therefore, the parametric analysis conducted using the mathematical model reveals that the two-tube and the annular fin configurations improve the dehumidification capacity under similar operating conditions.

**Table 4.3** Geometric description and parametric analysis methodology of the different fin-tube DCHE configurations

| Configuration | Description | Features |
|---|---|---|
| Single fin-tube: Fig. 4.5a | The existing heat exchanger configuration for which the mathematical model was validated | – |
| Two fin-tube: Fig. 4.5b | An additional cooling water source is added to the single fin-tube configuration | • The air and water flow occur in a crossflow arrangement<br>• Fin surface area, coating thickness, airflow rate, and water flow rate are similar to that of the single fin-tube configuration |
| Three fin-tube staggered: Fig. 4.5c | Three tubes are arranged in a staggered configuration and added to the same fin area of the single fin-tube configuration | |
| Annular fin-tube: Fig. 4.5d | Annular fins are attached to the tubes instead of the rectangular ones | • The heat transfer area of the rectangular fins and annular fin are identical<br>• A counter flow direction for air and water is assumed |
| No fin: Fig. 4.5e | No fins are considered, and the desiccant is coated directly on the copper tube | • The airflow occurs externally over the tubes while the water flows internally<br>• A counter flow arrangement is implemented<br>• The desiccant-air contact time is maintained similar to the single fin-tube configuration |

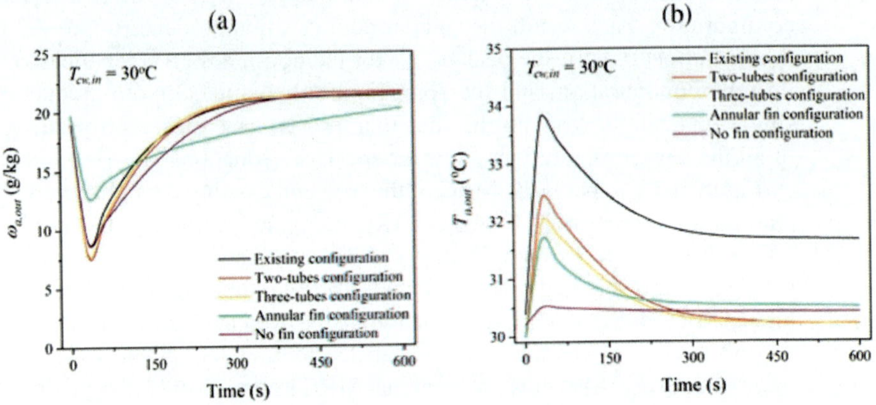


**Fig. 4.6** Dynamic response of outlet air humidity ratio and temperature of novel designs of DCHEs at cooling water temperature 30 °C

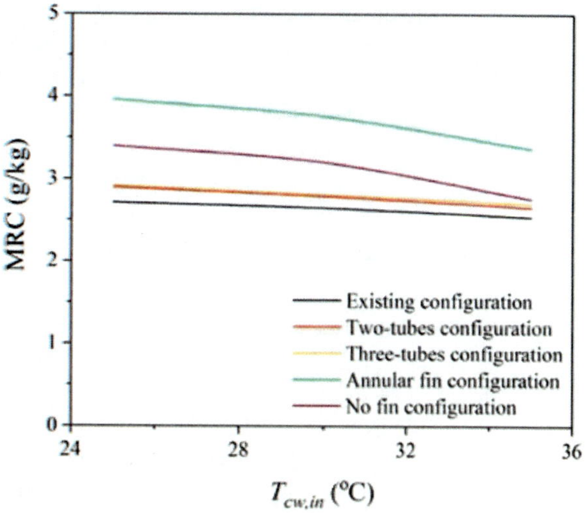


**Fig. 4.7** MRC of DCHE with different configuration as a function of cooling water temperature

## 4.3 Energy Analysis

Albeit the CFD model being able to predict the performance of DCHEs with reasonable accuracy, it does not directly compute the system-level energy savings realized by DCHEs for a specific application. Therefore, the first law of thermodynamics-based energy analysis is employed to study a hybrid cooling system incorporating the composite polymer DCHE and a conventional AHU coupled with an MVC chiller [14]. A cooling space with 1 refrigeration ton load (1RT ~3.517 kW) is considered, and the performance of DCHEs under different climate conditions is analyzed.

### *4.3.1 Assumptions*

The ensuing assumptions are made to simplify the analysis:

1. The indoor air temperature and humidity requirements are set as 24 °C and 50% RH based on the thermal comfort standards specified by ASHRAE [15]. The room return air conditions are chosen as 26 °C and 60% RH.
2. The standard COP of the MVC chiller is taken as 3 [16].
3. A cooling tower is used to supply the cooling water requirements of DCHE. Its efficiency and power requirement are selected as 75% and 0.04 kW/RT, respectively [17].
4. The DCHE's regeneration energy is supplied by integrating condenser/industrial waste heat or utilizing a solar thermal collector with around 60% overall efficiency [18].
5. The fan power consumption is governed by the pressure increase across the fan, volumetric airflow rate, pressure drop in the ducts, and fan shaft and motor

efficiencies. Stephens, Novoselac, and Siegel [19] established that the fan power consumption ranges between 18 and 22% of the total electricity consumption in a standalone MVC system. As a result, a correction factor of 1.2 is plugged to the equation while evaluating the overall electrical power consumption of the hybrid system.

### 4.3.2 Evaluation Methodology

Two commonly encountered configurations in a central air-conditioning system are considered, and the electrical power saved (EPS) by employing the hybrid system in these configurations is evaluated. In Fig. 4.8a, the hybrid system takes 100% fresh air, whereas, in Fig. 4.8b, 70% of the room return air is recirculated [20]. The DCHE system is placed upstream of the cooling coil to handle the fresh air. Four specific climatic regions that represent typical tropical climate are chosen for the analysis and are listed in Table 4.4.

### 4.3.3 Performance Parameters

The electrical power consumed by the chiller ($P_{el,ch}$) is computed as

$$P_{el,ch} = \frac{1.2\dot{m}_{a,cc}\left(h_{a,in,cc} - h_{a,out,cc}\right)}{\mathrm{COP}_{ch}} \tag{4.22}$$

where $\dot{m}_{a,cc}$ is the mass flow rate of air passing through the cooling coil; $h_{a,in,cc}$ and $h_{a,out,cc}$ are the respective inlet and outlet specific enthalpies of the air in the cooling coil; and $\mathrm{COP}_{ch}$ is the coefficient of performance of the chiller. The factor 1.2 accounts the contribution of fan power.

The electrical power needed for the cooling tower ($P_{el,\mathrm{CT}}$) is calculated by Eq. (4.23).

$$P_{el,\mathrm{CT}} = 0.0114\dot{m}_{a,\mathrm{DCHE}}\left(h_{a,in,\mathrm{DCHE}} - h_{a,out,\mathrm{DCHE}}\right) \tag{4.23}$$

where $\dot{m}_{a,\mathrm{DCHE}}$ is the mass flow rate of air passing through the DCHE and $h_{a,in,\mathrm{DCHE}}$ and $h_{a,out,\mathrm{DCHE}}$ are the respective inlet and outlet specific enthalpies of the air passing through the DCHE.

The electrical pumping power required for recirculating the water ($P_{el,sc}$) through the solar collector is computed via

$$P_{el,sc} = \frac{\dot{m}_{hw}\Delta P}{\rho_w \eta_{shaft} \eta_{motor}} \tag{4.24}$$

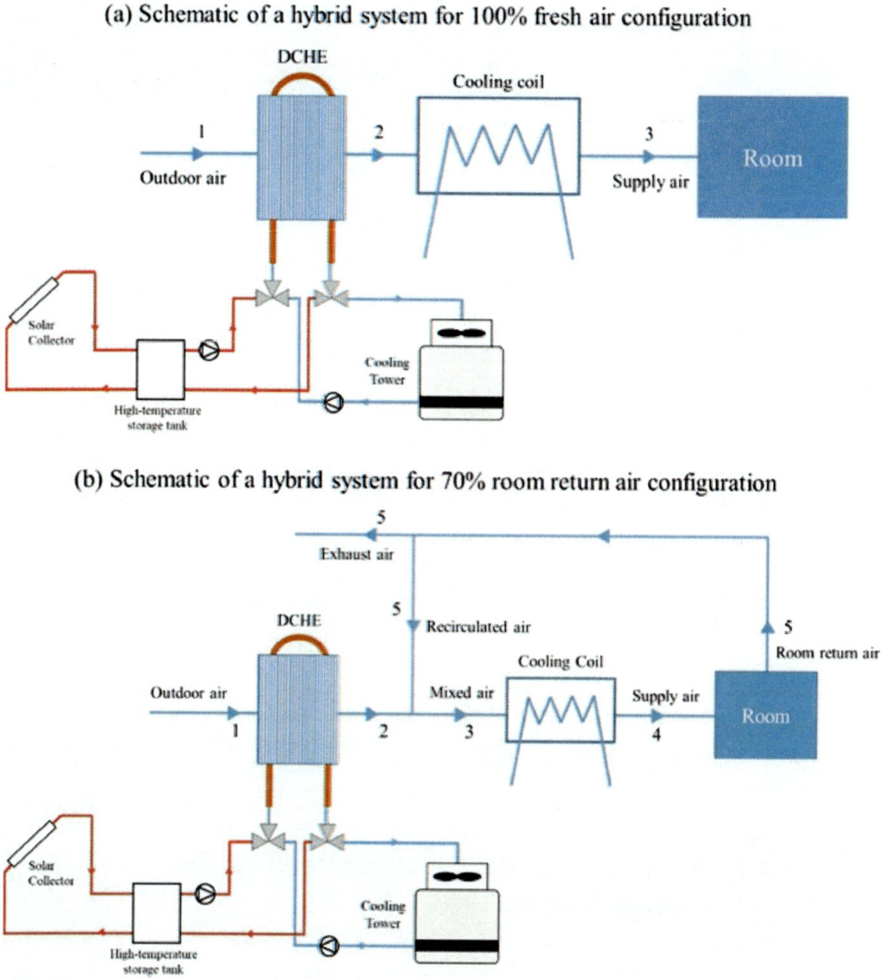


**Fig. 4.8** A schematic of the proposed hybrid central air-conditioning system comprising SAP-LiCl (50 w%) coated heat exchanger and conventional vapor compression chiller **a** for a 100% fresh air intake configuration and **b** with 70% room return air recirculation configuration

where $\dot{m}_{hw}$ is the regeneration water flow rate; $\rho_w$ is the density of liquid water; $\Delta P$ is the required pressure drop; $\eta_{shaft}$ and $\eta_{motor}$ are the efficiency of shaft and motor employed. An average pressure drop of around 25 kPa is assumed in the collector system [18], and the values of $\eta_{shaft}$ and $\eta_{motor}$ are taken as 0.6 and 0.7, respectively [19].

The electrical power saved (EPS) because of the employment of DCHEs is computed via the following equation:

$$\mathrm{EPS}(\%) = \frac{P_{el,ch} - \left(P'_{el,ch} + P_{el,\mathrm{CT}} + P_{el,sc} + P_{fan,DCHE}\right)}{P_{el,ch}} \times 100 \tag{4.25}$$

**Table 4.4** Four typical summer conditions in different geographical locations considered for energy analysis

| Case | City | Climatic condition | $T_{a,in}$ (°C) | RH (%) | $\omega_{a,in}$ (g/kg) | $I$(W/m$^2$) |
|---|---|---|---|---|---|---|
| A | Singapore | Moderate temperature and high humidity | 31 | 75 | 21.4 | 700 |
| B | Mumbai | High temperature and high humidity | 36 | 90 | 34.7 | 670 |
| C | Las Vegas | High temperature and moderate humidity | 40 | 40 | 18.7 | 690 |
| D | New York | Moderate temperature and moderate humidity | 30 | 40 | 11.3 | 570 |

where $P'_{el,ch}$ is the electrical power consumed by the chiller in the AHU containing single/multiple DCHE stages, and $P_{fan,DCHE}$ is the excess fan power consumed by the DCHE and is taken as 10% of the cooling load removed by the DCHE.

The area occupied by the solar collector is computed by

$$A_{sc} = \frac{\dot{m}_{a,\mathrm{DCHE}}\left(\omega_{a,in,\mathrm{DCHE}} - \omega_{a,out,\mathrm{DCHE}}\right)\Delta H_s}{\eta_r \eta_{sc} I} \tag{4.26}$$

where $\eta_{sc}$ is the efficiency of the solar collector and $I$ is the solar irradiance. The solar collector efficiency value of 0.6 is consistently adopted in the analysis.

### 4.3.4 *Effect of Climatic Conditions and Airflow Configurations*

The energy savings potential of the hybrid cooling system is determined under different climatic conditions. The operating parameters for the system-level performance studies are listed in Table 4.5. The transient $\omega_{a,out}$ and $T_{a,out}$ profiles are displayed in Fig. 4.9 and their time-averaged values are highlighted in Table 4.6.

**Table 4.5** Operating parameters of SLCHE for the system-level energy analysis

| Case index | $T_{a,in}$ (°C) | $\omega_{a,in}$ (g/kg) | $T_{cw,in}$ (°C) | $T_{hw,in}$ (°C) | $\dot{m}_a$ (kg/h) | $t_{cyc}$ (min) | $\dot{m}_w$ (kg/min) |
|---|---|---|---|---|---|---|---|
| A | 30 | 21.4 | 28.4 | 50 | 35 | 10 | 4 |
| B | 36 | 34.7 | 25.7 | | | | |
| C | 40 | 18.7 | 30.3 | | | | |
| D | 30 | 11.3 | 23.3 | | | | |

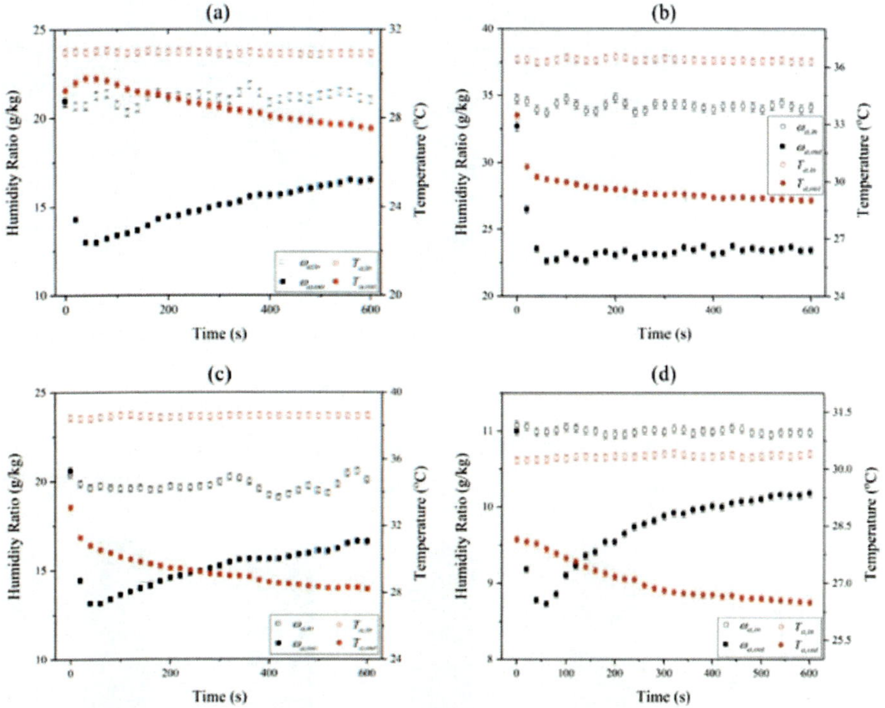


Fig. 4.9 Dynamic behavior of inlet and outlet air humidity ratio and temperature of SLCHE under **a** Case A: moderate temperature and high humidity conditions of 31 °C and 75% RH; **b** Case B: high temperature and high humidity conditions of 36 °C and 90% RH; **c** Case C: high temperature and moderate humidity conditions of 40 °C and 40% RH; and **d** Case D: moderate temperature and moderate humidity conditions of 30 °C and 40% RH

**Table 4.6** Time-averaged values of air temperature and humidity ratio at the DCHE outlet for different climate conditions

| Case index | $T_{a,in}$ (°C) | $\omega_{a,in}$ (g/kg) | $\bar{T}_{a,out}$ (°C) | $\bar{\omega}_{a,out}$ (g/kg) |
|---|---|---|---|---|
| A | 30 | 21.4 | 29.7 | 15.5 |
| B | 36 | 34.7 | 29.3 | 23.2 |
| C | 40 | 18.7 | 28.4 | 14.3 |
| D | 30 | 11.3 | 27.1 | 9.8 |

DCHE reduces the air temperature by 2–30% and dehumidifies it by 15–40%. The resultant reduction in the cooling load and EPS are evaluated using Eqs. (4.22)–(4.25) and are as shown in Fig. 4.10. DCHE handles 32–75% of the cooling load operating under 100% fresh air configuration and correspondingly slashes the electricity consumption by 40–50%. Specifically, for cases A and B, the DCHE functions as a dehumidifier and lowers the latent load in the air by 50%, while marginally (<5%)

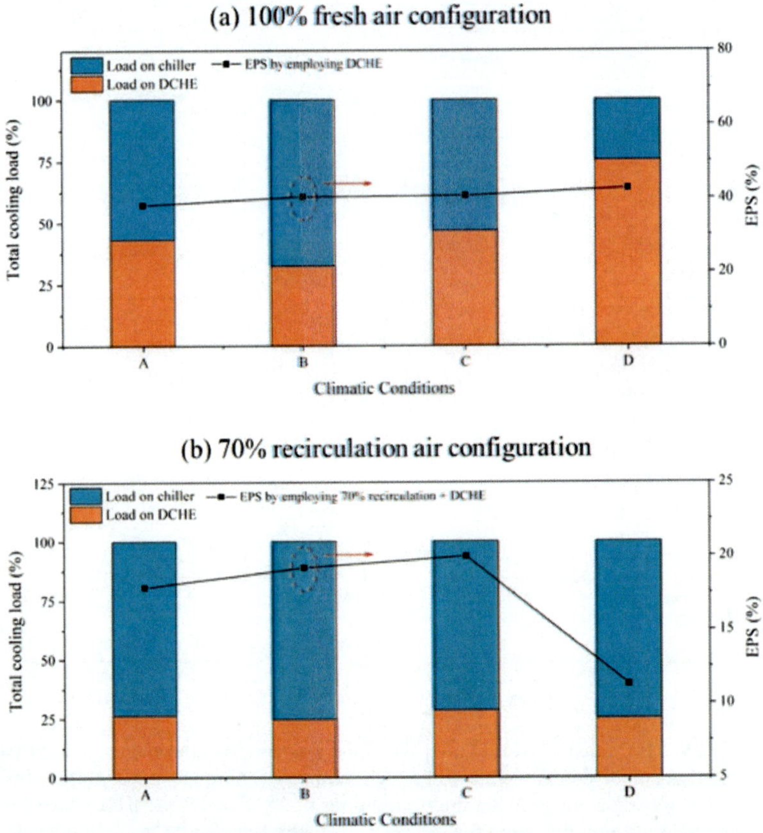


**Fig. 4.10** Total cooling load reduction and electrical power saved (EPS) by employing the hybrid air-conditioning system for different climatic environments under **a** fully fresh-air configuration; and **b** 70% recirculated air configuration

reducing the sensible load. In contrast, for cases C and D, both sensible and latent loads are markedly reduced by 40%.

When 70% of the return air is being recirculated, the cooling load can be trimmed by 15–50% because of the adiabatic mixing process (state 2 and 5 mix to form state 3 in Fig. 4.8b). By employing a DCHE in this configuration, an additional 20–30% of the chiller cooling load is reduced, which leads to electrical power reduction of about 11–20%. Moreover, the employment of DCHEs paves the way for increasing the cooling coil temperature by 8–12 °C. Figure 4.11 illustrates that the chiller's COP improves by about 3.7% for every 1 °C rise in chilled water temperature [21, 22]. Lastly, Fig. 4.12 depicts the solar collector area required for regenerating DCHEs. The collector footprint can be trimmed by 65–75% when the fresh air requirement is confined to 30%. In sum, the energy analysis shows the performance improvement and energy benefits yielded by the DCHE technology under varying ambient conditions.

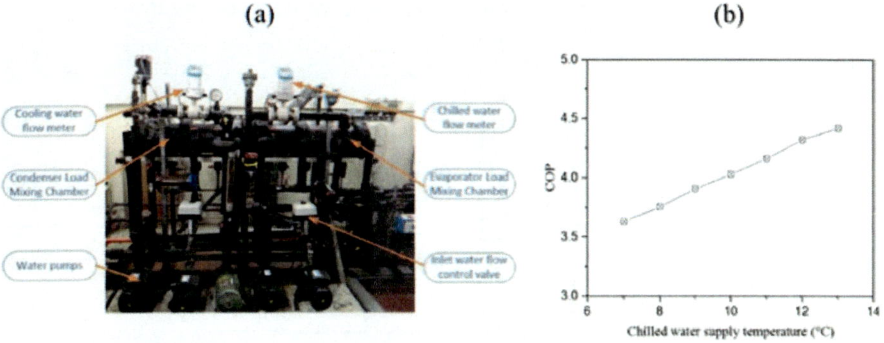


**Fig. 4.11** **a** Experimental setup to determine the impact of chiller water supply temperature [21]; and **b** the effect of chiller water supply temperature on the COP of the chiller system [22]

**Fig. 4.12** Solar collector area under different climatic conditions for fully fresh air and 70% recirculated air configurations of the hybrid air-conditioning system

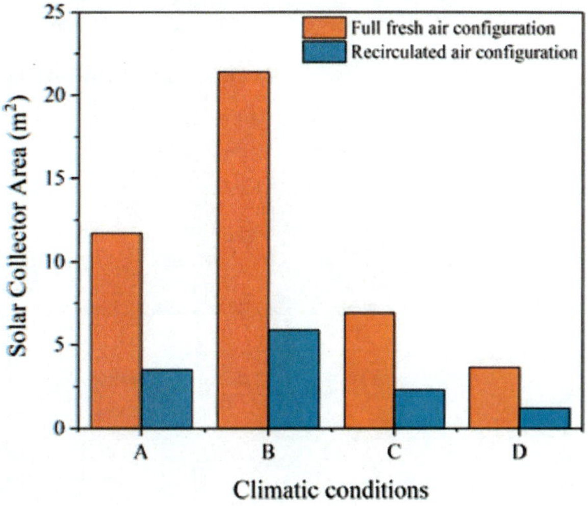


## 4.4 Economic Analysis

The cost savings potential offered by the hybrid cooling system is computed in this section. Figure 4.13 presents four new cases of the hybrid air-conditioning system considered for this analysis, including a single-stage DCHE with/without cooling tower and two-stage DCHE with/without cooling tower. The surplus capital expenditure ($CAPEX_{surplus}$) required for the hybrid system and the corresponding annual operating expenses saved ($OPEX_{saved}$) are evaluated. The results are compared with the standalone MVC system to compute the payback period and rate of return [23].

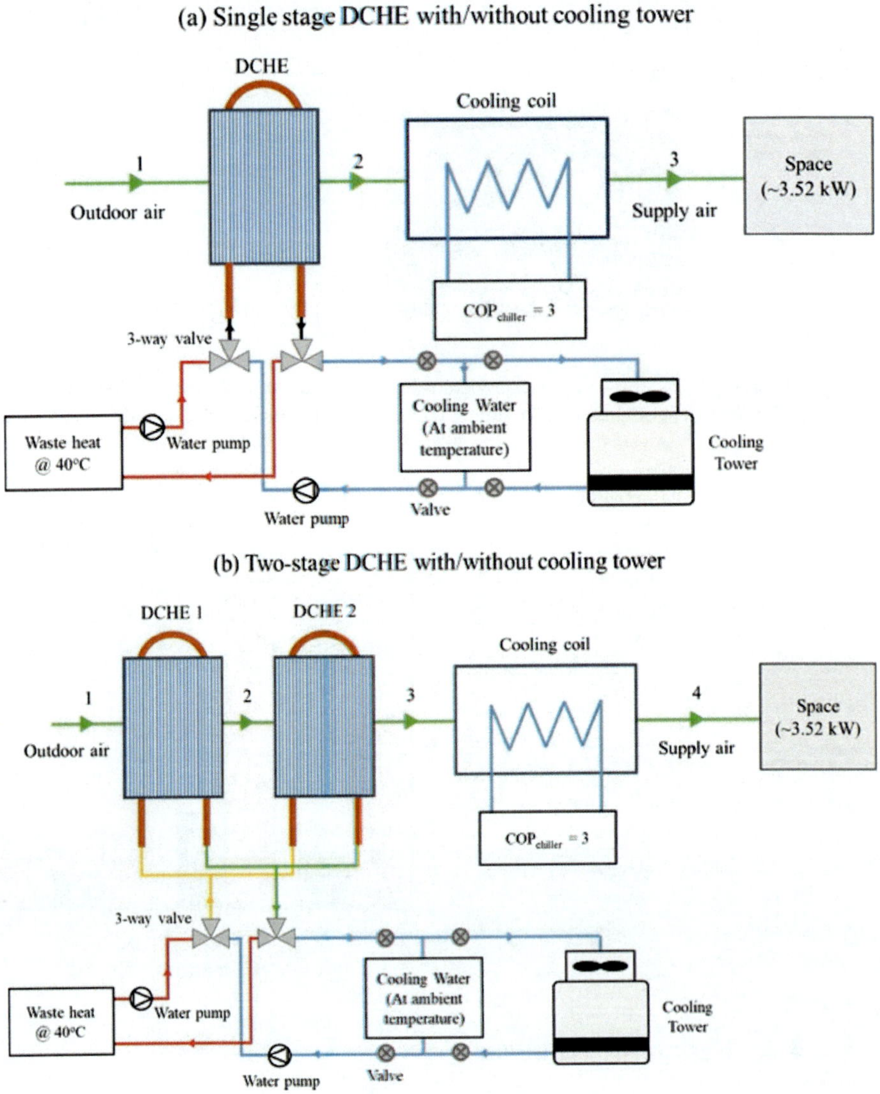


**Fig. 4.13** Schematic of **a** single-stage DCHE configuration and **b** two-stage DCHE configuration in an AHU. It is noteworthy that the DCHEs are regenerated with waste heat at 40 °C and provided with cooling water at ambient temperature or supplied through a cooling tower

### *4.4.1 Surplus Capital Expenditure*

In this analysis, the DCHE system is assumed to retrofit in the AHU along with the existing chiller and cooling coil. As a result, the capital cost of the central chiller is not accounted for in the computation. A fin-tube heat exchanger's fabrication cost sums up to US$ 500 equivalent [24]. In addition to this cost, the capital expenditure for DCHE also entails the expense incurred to coat the desiccant material on the heat exchanger. The superabsorbent polymer and hygroscopic salt are assumed to cost about US$ 140 per kg, respectively [25, 26]. Suppose 1.5 kg of the composite desiccant is necessary to achieve a 100 $\mu$m coating thickness. In that case, the cost of preparing the desiccant solution sums to US$ 300 equivalent, including the material cost and wastage. In addition, the heat exchanger must be dried to achieve a dry desiccant layer on the heat exchanger surface. To compute this term, the electricity expenditure to dry the heat exchanger is required. The cost of electricity (COE) is taken as US$ 0.18 equivalent per kWh, and it is assumed to inflate annually by 2% [27]. If the heat exchanger is dipped 10 times in the desiccant solution to achieve the required desiccant coating thickness or mass, and a drying time of 2 h per coating is necessary for a 5 kW rated industrial oven [28], the electricity cost amounts to US$ 20. In addition, if a cooling tower is employed, its cost is evaluated using Eq. (4.27) [29].

$$\text{CAPEX}_{\text{CT}} = \text{US\$}250\left(\dot{Q}_{DCHE}/3.52\right) \tag{4.27}$$

where $\text{CAPEX}_{\text{CT}}$ is the capital expense of the cooling tower and $\dot{Q}_{DCHE}$ is the total cooling load removed by the DCHE.

The surplus capital expenditure $\left(\text{CAPEX}_{\text{surplus}}\right)$ involved in switching to the hybrid cooling system is evaluated using

$$\text{CAPEX}_{\text{surplus}} = g(\text{CAPEX}_{\text{HX}} + \text{Cost of coating}) + \text{CAPEX}_{\text{CT}} + \text{Retrofitting Cost} \tag{4.28}$$

where $g$ is the number of DCHE stages employed in the system.

### 4.4.2 Operating Expenditure

The yearly operating expenditure comprises the costs required for operating and maintaining the system which is the net cash flow of the system for 5 years. While evaluating the operating cost is straightforward based on the hybrid system's operation strategy, estimating its maintenance cost is complex. Accordingly, 10% of savings realized in the operating cost is discounted as the cost of maintenance.

The yearly expenditure associated with operating the conventional chiller system $(\text{OPEX}_{\text{conventional}})$ is evaluated using

$$\text{OPEX}_{\text{conventional}} = \text{EC} \times P_{el,ch} \times \tau \times 365 \tag{4.29}$$

where EC is the electricity cost; $\tau$ is the number of hours of operation per day and is taken as 8 h in this study for consistency; and $P_{el,ch}$ is the electrical power of the conventional chiller as evaluated by Eq. (4.22).

Similarly, the annual cost incurred to run the hybrid air-conditioning system $(\text{OPEX}_{\text{hybrid}})$ is computed as

$$\text{OPEX}_{\text{hybrid}} = \text{COE} \times \left(P'_{el,ch} + P_{CT}\right) \times \tau \times 365 \tag{4.30}$$

where $P'_{el,ch}$ is the electrical power of the chiller in the hybrid system and $P_{CT}$ is the cooling tower electrical power consumption. The pumping power consumption for providing the cooling water at ambient temperature is evaluated by considering the power consumption of a small cooling tower that can supply water at the ambient temperature.

### 4.4.3 Performance Parameters

The annual savings in operating expenses is computed via

$$\text{OPEX}_{\text{saved}} = \text{OPEX}_{\text{conventional}} - \text{OPEX}_{\text{hybrid}} \tag{4.31}$$

In a cash flow analysis, the $\text{OPEX}_{\text{saved}}$ is considered as the yearly cash inflow whereas the maintenance cost, evaluated at 10% of the $\text{OPEX}_{\text{saved}}$ is counted as the yearly outflow. The annual net cash flow is then evaluated using Eq. (4.32), where $NCF_z$ is the net cash flow in the year z.

$$NCF_z = \text{Cash inflow} - \text{Cash outflow} \tag{4.32}$$

Cumulative cash flow is computed by summing up all the yearly cash flows including the initial capital cost.

$$CCF_n = \text{CAPEX}_{\text{surplus}} + \sum_{z=1}^{n} NCF_z \tag{4.33}$$

where $CCF_n$ is the cumulative cash flow at the end of nth year, which is taken as 5 for this analysis.

The internal rate of return (IRR) that equates the value of the future cash flows to the initial capital cost, is computed via

$$CCF_n = \sum_{z=0}^{n} \frac{NCF_z}{(1 + IRR)^z} \quad (4.34)$$

### 4.4.4 Effect of DCHEs on Cost Savings

Figure 4.14 portrays the cash flow analysis of the hybrid system configurations. It illustrates the $CAPEX_{surplus}$ required to retrofit DCHEs and then compares the $OPEX_{saved}$. The cost of installing a cooling tower in the system is only 50% higher than the DCHE system without a cooling tower. However, if two DCHE stages are desired, then the system incurs two times higher capital costs. Despite the rise in $CAPEX_{surplus}$, the additional DCHE stage improves the $OPEX_{saved}$ by around two times, and the cooling tower reduces the operating cost by 50%. The improvement in $OPEX_{saved}$ is primarily due to higher EPS, as portrayed in Fig. 4.15a. Therefore,

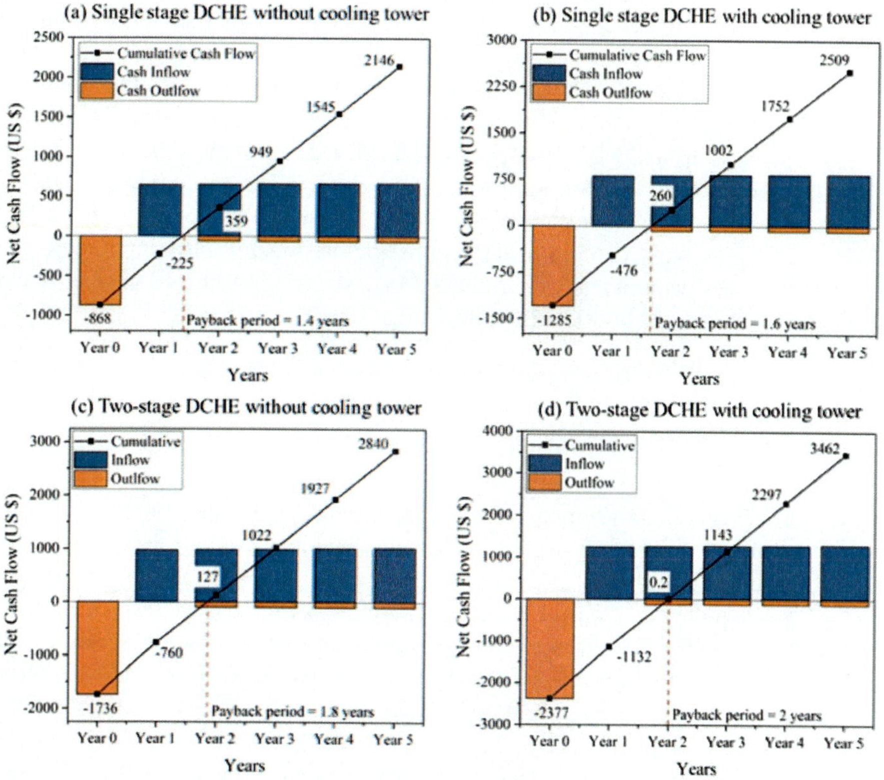


**Fig. 4.14** Net Cash Flow analysis of **a** Single-stage DCHE without cooling tower configuration; **b** Single-stage DCHE with cooling tower configuration; **c** Two-stage DCHE without cooling tower; and **d** Two-stage DCHE with cooling tower configuration

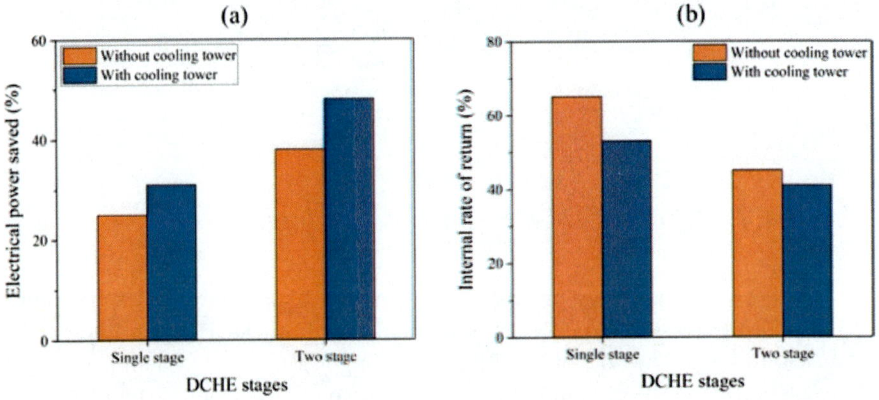


**Fig. 4.15** **a** EPS recorded by single and two-stage DCHEs with/without the presence of the cooling tower; and **b** Internal rate of return for the four configurations of hybrid air conditioner

shorter payback periods of <2 years can be realized, and a positive cash flow is observed from Year 2 onwards. Since the "single DCHE stage without cooling tower" configuration generates positive cash flow more rapidly than other configurations, it provides the highest returns of around 65%, as shown in Fig. 4.15b. However, the "two-stage DCHE with cooling tower" configuration exceeds the amount saved by the single-stage configuration by around 12% from Year 3 onwards, and the savings increase up to 21% at the end of Year 5. Due to higher $CAPEX_{surplus}$, the "two-stage DCHE with cooling tower" configuration yields the lowest IRR. However, it records 60% higher cumulative cash flow when compared to the "single-stage DCHE without cooling tower" configuration. In sum, this simplified cost analysis approach demonstrates the remarkable cost savings potential gained by employing a DCHE-based hybrid air-conditioning system.

## 4.5 Exergy Analysis

The first law analysis quantifies the cooling and regeneration energy inputs against the cooling load of the system. However, it is unable to evaluate the quality of the energy required nor does it explain the causes of the DCHE system's irreversibility. It is worth noting that the second law of thermodynamics imposes a process-independent lower limit on the energy consumption of the air-conditioning process. This is the absolute minimum amount of energy necessary for operating an ideal air-conditioner. In comparison to this ideal device, the amount of entropy produced by the DCHE system must be computed. As a result, the second law of thermodynamics method is used in this section to calculate the irreversibility associated with DCHE's operation [30]. DCHE is modeled as a thermodynamic system, and its available/useful energy is computed with reference to a pre-determined dead state. The selection of the dead state is essential as it controls the extent of the exergy in the system. Since this is an air-conditioning application, the ambient state is considered as the dead state [31]. It is worth noting that the fundamental definitions of the second law of thermodynamics, namely, exergy, entropy, exergy balance, entropy generation, dead state, available energy, etc., are available for review in [32, 33].

### 4.5.1 Thermodynamic Least Work for Air-Conditioning Process

Figure 4.16 shows the schematic of an ideal air-conditioner and its reversible thermodynamic paths on a psychrometric chart. In the first stage, the air is dehumidified in a quasi-steady reversible isothermal process. Its work input is calculated by applying the exergy balance approach to its reverse process, i.e., reversible isothermal mixing of air and water vapor [34]. The amount of exergy destroyed in the reversible isothermal mixing process is considered as the least work input for the reversible isothermal dehumidification process, as shown in Eq. (4.35)

$$\dot{W}_{rev,1-2'} = \dot{m}_a\left(\xi_{a,2'} + \left(\frac{\omega_1 - \omega_2}{1000}\right)\xi_{w,2'} - \xi_{a,1}\right) \tag{4.35}$$

where $\dot{m}_a$ is the mass flow rate of dry air; $\omega$ is the humidity ratio of air; $\xi_a$ is the specific flow exergy of moist air; and $\xi_w$ is the specific flow exergy of liquid water.

The moist air is assumed to be an ideal mixture of dry air and water vapor, and the thermodynamic properties of liquid water are taken as that of a saturated liquid at the desired temperature. The specific flow exergy of moist air and water is

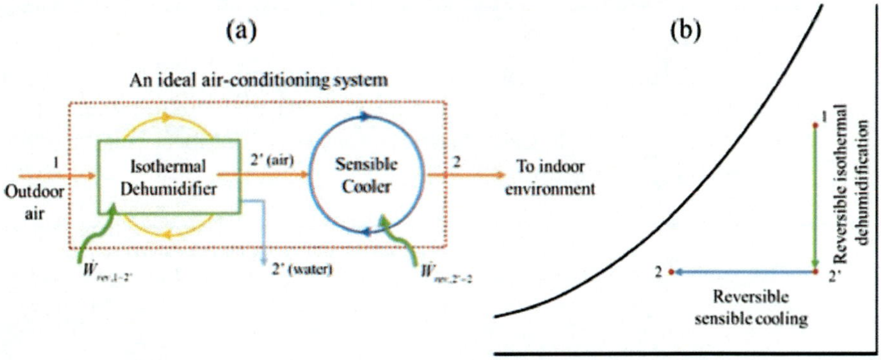


**Fig. 4.16** Schematic of **a** an ideal air-conditioner for calculating the thermodynamic least work; and **b** psychrometric chart highlighting the two stages of the thermodynamic paths involved in the ideal air-conditioner

then evaluated by employing Eqs. (4.36) and (4.37), respectively [33]. The thermo-mechanical components of specific flow exergy quantify the extent to which the system's temperature and pressure differ from the dead state. The chemical exergy denotes the change in the chemical composition of the system with respect to its dead state.

$$\xi_a = \underbrace{\left(c_{p,a} + \omega c_{p,v}\right) T_0 \left[\frac{T}{T_0} - 1 - \ln\left(\frac{T}{T_0}\right)\right]}_{\text{Thermal component}} + \underbrace{(1 + \tilde{\omega}) R_a T_0 \ln\left(\frac{P}{P_0}\right)}_{\text{Mechanical component}}$$
$$+ \underbrace{R_a T_0 \left[\ln\left(\frac{1 + \tilde{\omega}_0}{1 + \tilde{\omega}}\right) + \tilde{\omega} \ln\left(\frac{\tilde{\omega}}{\tilde{\omega}_0} \frac{1 + \tilde{\omega}_0}{1 + \tilde{\omega}}\right)\right]}_{\text{Chemical component}} \tag{4.36}$$

$$\xi_w = \underbrace{h_f(T) - h_g(T_0) - T_0\left(s_f(T) - s_g(T_0)\right)}_{\text{Thermal component}} + \underbrace{(P - P_{sat}(T)) v_f(T)}_{\text{Mechanical component}}$$
$$- \underbrace{R_v T_0 \ln \varphi_0}_{\text{Chemical component}} \tag{4.37}$$

where $c_{p,a}$ and $c_{p,v}$ are the respective specific heat capacities of dry air and water vapor at constant pressure; $T$ is the absolute temperature; $T_0$ is the dead state temperature; $\tilde{\omega}$ is the mole fraction ratio of vapor to air with $\tilde{\omega} = 1.608\omega$; $R_a$ is the gas constant of dry air; $h_f(T)$ and $s_f(T)$ are the respective specific enthalpy and entropy values of water in the saturated liquid state evaluated at $T$; $h_g(T_0)$ and $s_g(T_0)$ are the respective specific enthalpy and entropy values of water in the saturated vapor state evaluated at $T_0$; $P$ is the pressure of air (taken as $P_{atm} = 101.325$ kPa in this study); $P_{sat}$ is the saturation pressure of water evaluated at $T$; $R_v$ is the gas constant for water vapor; and $\varphi_0$ is the ambient relative humidity ratio. The mathematical correlations for the thermodynamic state properties of water were directly adopted from Saul and Wagner [35].

In the second stage, the dehumidified air is sensibly cooled without any change in its chemical composition. The minimum work required for this process is evolved by considering a reversible Carnot refrigerator exchanging heat with the environment [34] and is represented as

$$\dot{W}_{rev,2'-2} = \int_{T_{2'}}^{T_2} \left(1 - \frac{T_0}{T}\right) \delta Q = \dot{m}_a \left(c_{p,a} + \frac{\omega_2}{1000} c_{p,v}\right) \left[T_0 \ln \frac{T_2}{T_{2'}} - (T_2 - T_{2'})\right] \tag{4.38}$$

Accordingly, the least work input for the air-conditioning process is determined as

$$\dot{W}_{rev,1-2} = \dot{W}_{rev,1-2'} + \dot{W}_{rev,2'-2} \tag{4.39}$$

### 4.5.2 Actual Work in DCHEs

Figure 4.17 portrays the control volume of a DCHE. A steady-state control volume analysis of the first law of thermodynamics is applied to obtain the energy balance of DCHE during dehumidification and regeneration processes and is respectively shown in Eqs. (4.40) and (4.41).

$$\dot{m}_{a,d}\left(h_{a,d,in} - h_{a,d,out}\right) + \dot{m}_{cw}c_{p,w}\left(T_{cw,in} - T_{cw,out}\right) - \dot{Q}_{lost,d} = 0 \tag{4.40}$$

$$\dot{m}_{a,r}\left(h_{a,r,in} - h_{a,r,out}\right) + \dot{m}_{hw}c_{p,w}\left(T_{hw,in} - T_{hw,out}\right) - \dot{Q}_{lost,r} = 0 \tag{4.41}$$

where $\dot{m}_{a,d}$ and $\dot{m}_{a,r}$ are the respective mass flow rates of air during dehumidification and regeneration processes; $h_{a,d,in}$, $h_{a,d,out}$, $h_{a,r,in}$, and $h_{a,r,out}$ are the specific enthalpies of moist air at the inlet and outlet of dehumidification and regeneration processes, respectively; $\dot{m}_{cw}$ and $\dot{m}_{hw}$ represent the respective mass flow rates of cooling water and hot water; $c_{p,w}$ is the specific heat capacity at constant pressure for liquid water; $T_{cw,in}/T_{hw,in}$ and $T_{cw,out}/T_{hw,out}$ are the respective cooling/hot water temperature at DCHE inlet and outlet; and $\dot{Q}_{lost,d}$ and $\dot{Q}_{lost,r}$ are the respective heat loss rates of the DCHE during dehumidification and regeneration processes.

The specific enthalpy of moist air ($h_a$) is defined as the total enthalpy of dry air and water vapor mixture per every kg of dry air [36] and is computed via

$$h_a = c_{p,a}T + \omega\left(c_{p,v}T + 2501\right) \tag{4.42}$$

where $T$ is the air temperature and $\omega$ is the specific humidity of moisture in air.

The overall exergy balance associated with the operation of DCHE under dehumidification and regeneration processes is carried out via Eqs. (4.43) and (4.44),

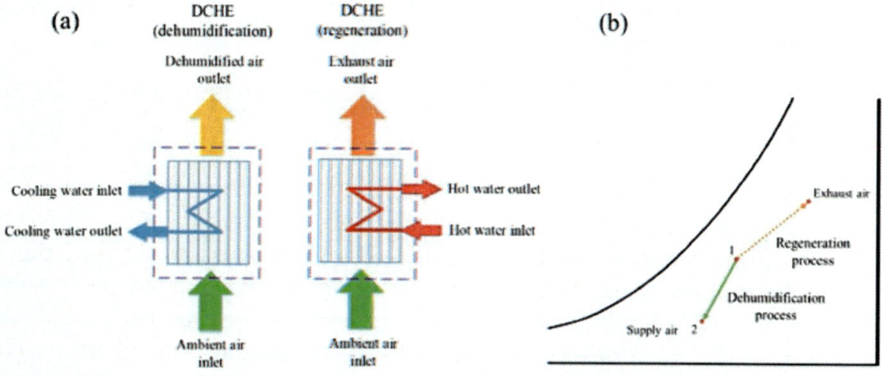


**Fig. 4.17** Schematic of **a** a DCHE system undergoing dehumidification and regeneration processes; and **b** psychrometric chart highlighting the corresponding thermodynamic paths

respectively. Exergy balance analysis combines both the first and second laws of thermodynamics to assess the quality of the corresponding energy transfer. Lior and Al-Sharqawi [37] concluded that the average exergy of the humid air stream dominates over the total exergy in the desiccant. Therefore, the contribution from the desiccant toward the exergy is taken as negligible.

$$\begin{aligned}\dot{E}x_{dest,d} &= \dot{m}_{a,d}\left(\xi_{a,d,in} - \xi_{a,d,out}\right) + \dot{m}_{cw}\left(\xi_{cw,in} - \xi_{cw,out}\right) \\ &\quad - \dot{m}_{a,d}\left(\omega_{a,d,in} - \omega_{a,d,out}\right)\xi_{ad,water} - \dot{Q}_{lost,d}\left(1 - \frac{T_0}{T_d}\right)\end{aligned} \tag{4.43}$$

$$\begin{aligned}\dot{E}x_{dest,r} &= \dot{m}_{a,r}\left(\xi_{a,r,in} - \xi_{a,r,out}\right) + \dot{m}_{hw}\left(\xi_{hw,in} - \xi_{hw,out}\right) \\ &\quad + \dot{m}_{a,r}\left(\omega_{a,r,out} - \omega_{a,r,in}\right)\xi_{des,water} - \dot{Q}_{lost,r}\left(1 - \frac{T_0}{T_r}\right)\end{aligned} \tag{4.44}$$

where $\dot{E}x_{dest,d}$ and $\dot{E}x_{dest,r}$ are the exergy destruction rates associated with the DCHE's operation during dehumidification and regeneration modes; and $T_0$ is the absolute dead state temperature. The first two terms on the right side of Eqs. (4.43) and (4.44) represent the exergy associated with the flow of moist air and cooling water/hot water during dehumidification and regeneration processes. The third term indicates the exergy storage in the system as the result of sorption/desorption of moisture. Further, the last term in both the equations represents the exergy associated with the heat loss during dehumidification/regeneration process.

The absolute rate of entropy generation $\left(\dot{S}_{gen}\right)$ is obtained by combining the exergy destroyed during dehumidification and regeneration process and is defined as

$$\dot{S}_{gen} = \frac{\dot{E}x_{dest,d} + \dot{E}x_{dest,r}}{T_0} \tag{4.45}$$

Further, by employing the Gouy Stodola theorem [32], the relationship between the actual work input required for DCHE's operation and its thermodynamic lower limit is established as

$$\dot{W}_{actual} = \dot{W}_{rev} + T_0\dot{S}_{gen} \tag{4.46}$$

$\dot{S}_{gen}$, computed using Eq. (4.45), considers the overall entropy production rate during dehumidification and regeneration processes. However, it does not account for the moisture removal rate of the DCHEs. When comparing the entropy generated by different desiccants, $\dot{S}_{gen}$ changes due to the irreversibility in the system and the moisture quantity removed. To facilitate a proper comparison due to the irreversibility alone, $\dot{S}_{gen}$ is normalized with the desiccant's moisture removal rate. Accordingly, specific entropy generation ($s_g$) is introduced in Eq. (4.47), and is defined as the ratio of the entropy generated per unit mass of moisture removed from the air.

$$s_g = \frac{\dot{S}_{gen}}{\dot{m}_a\left(\omega_{a,in} - \omega_{a,out}\right)} \tag{4.47}$$

Another key indicator, the second law efficiency ($\eta_{II}$) is introduced to quantify the thermodynamic reversibility of DCHEs. It is defined as the ratio of the work input of an ideal air-conditioning system to the actual work input of the DCHE system.

$$\eta_{II} = \frac{\dot{W}_{rev}}{\dot{W}_{actual}} \tag{4.48}$$

### 4.5.3 Exergy Flow in DCHEs

Table 4.2 lists the operating parameters employed to study the second law performance of DCHEs. The testing facility described in Sect. 4.1.1 of Chap. 3 is used for experimental comparison. During both dehumidification and regeneration processes, the inlet air temperature and humidity ratio are maintained at 30 °C and 21.5 g/kg.

Figure 4.18a shows the time-averaged inlet and outlet states of humidity ratio, air, water, and desiccant surface temperatures. Using these experimentally measured temperature and humidity ratios, the total flow exergy of DCHEs is evaluated, as shown in Fig. 4.18b. Since the ambient conditions are employed as the dead state, the inlet air supply exergy is 0 W. This is because the air is in thermal, mechanical, and chemical equilibrium with the dead state. From our experiments, we measured that the pressure drop in the DCHEs marginally fluctuates between 10 and 15 Pa ($<$0.001% compared to $P_0 = 101.325$ kPa) for the selected range of airflow rate [38]. Therefore, the mechanical component of the specific flow exergy of air is considered negligible compared to thermal and chemical components.

The exergy analysis reveals that 3.694 W of net exergy is supplied through the cooling water during the dehumidification process. Since the air temperature and humidity ratio are reduced, the exergy of the outlet air rises to 1.157 W while the moisture captured by the desiccant reduces the system's exergy by 2.33 W. In sum, the dehumidification process destroys only 0.004 W, demonstrating excellent reversibility. In contrast, around 24.42 W of net exergy input is supplied via the hot water injection during the regeneration process and 17.06 W is destroyed. The regeneration process destroys higher exergy than the dehumidification process. The irreversibility occurs due to greater temperature difference between the hot water and the dead state and the limited heat transfer coefficient between hot water and air. Further, by employing Eq. (4.39), an insightful comparison between the DCHE system and an ideal reversible air-conditioner is made. The theoretical least work input is computed as 3.49 W, which marks the second law efficiency of the DCHE system as 17%. A low value of second law efficiency suggests that there is still scope

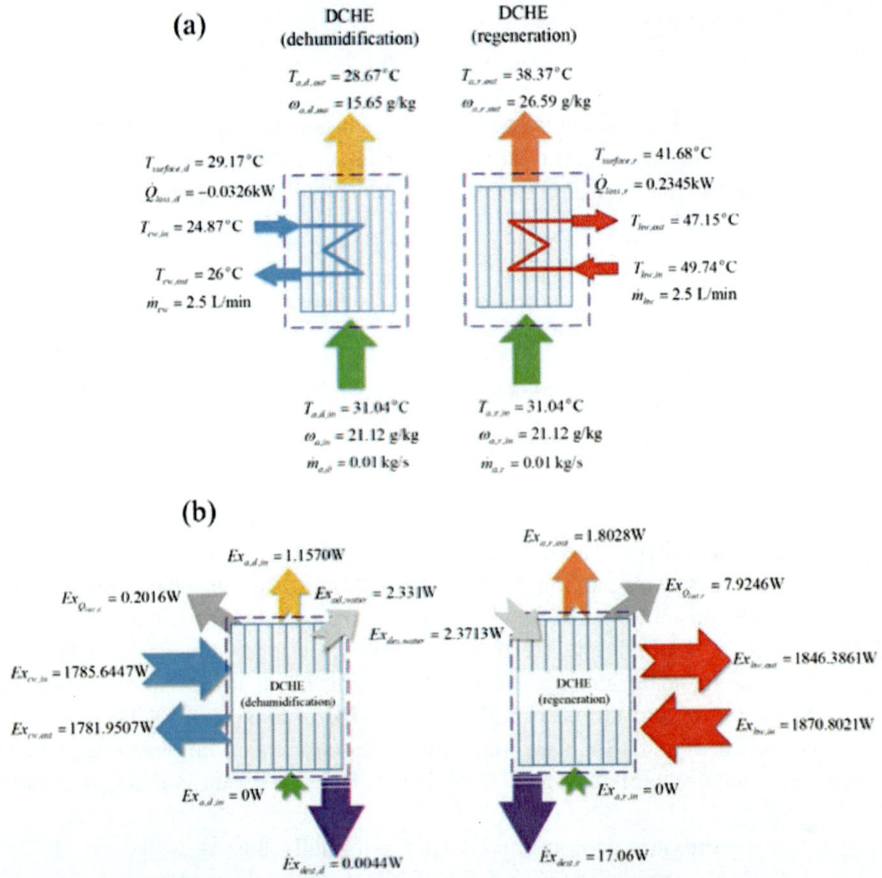


**Fig. 4.18** **a** Time-averaged cyclic inlet and outlet states of air temperature, air humidity, water temperature, and desiccant surface temperature; **b** exergy flow diagram in the simultaneous heat and mass transfer process of DCHEs

for performance improvement in DCHEs, and one of the possible ways is to reduce the regeneration temperature.

### *4.5.4 Effect of reducing regeneration temperature*

Figure 4.19 illustrates the trends in performance parameters under the influence of varying hot water temperature $\left(T_{hw,in}\right)$. The desiccant's MRC increases by about 18% from 4.3 to 5.1 g/kg when $T_{hw,in}$ is regulated between 40 and 70 °C. This is because higher $T_{hw,in}$ adequately regenerates the desiccant and removes any moisture trapping. However, the heat losses to the environment appreciate at higher $T_{hw,in}$, which trims the first law energy efficiency parameter, i.e., $COP_{th}$, by 40%. The first

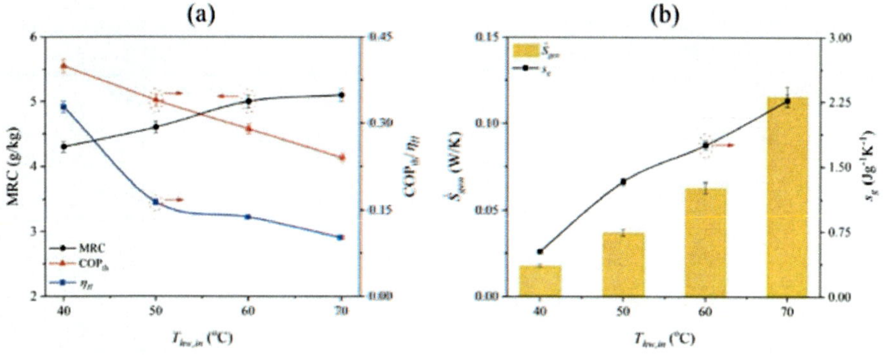


**Fig. 4.19** Effect of variation of hot water temperature $(T_{hw,in})$ from 40 to 70 °C on **a** MRC, $COP_{th}$, and $\eta_{II}$; and **b** $\dot{S}_{gen}$ and $s_g$

law of thermodynamics highlights the necessity of a compromise between improved dehumidification performance and energy efficiency.

The second law analysis shows that a higher $T_{hw,in}$ raises the temperature difference with the dead state. A reduction in the dehumidification capacity also modifies the chemical component of exergy. Accordingly, the entropy generation rates (both absolute and specific): $\dot{S}_{gen}$ and $s_g$, increase by six times and 300%, respectively, and $\eta_{II}$ drops by 70%. A close observation in the trend of $\eta_{II}$ reveals that around 75% drop occurs between 40 and 50 °C. This conveys higher sensitivity of entropy production rates in the temperature range spanning 40–50 °C. As a result, maintaining the regeneration temperature at 40 °C improves the DCHE system's potential to do useful work.

### *4.5.5 Effect of Desiccant Type*

The thermodynamic performance is evaluated for different desiccant materials in this section. Four DCHEs coated with different desiccants are considered for this study: silica gel, PVA-LiCl (50 w%), SAP-$HCO_2K$ (50 w%), and SAP-LiCl (50 w%). To ensure that the performance variation is purely due to the desiccant type employed, identical heat exchangers are used, and the amount of desiccant-coated is maintained around 50 g. The operating conditions were maintained stable, as listed in Table 4.7. Figure 4.20 depicts the desiccant's influence on the first law performance indicators, i.e., MRC and $COP_{th}$. Also, the effect on air temperature is also depicted. The composite polymer-based desiccants record around three times higher MRC compared to silica gel, while showing marginal effect on the outlet air temperature. The improvement in the dehumidification capacity also translates to higher first law efficiency. The composite polymer desiccants record over two times higher $COP_{th}$ than silica gel.

**Table 4.7** Experimental conditions for analyzing the second law performance of DCHEs

| Parameters | Units | Experimental conditions |
|---|---|---|
| Cooling water temperature ($T_{cw,in}$) | °C | 25 |
| Hot water temperature ($T_{hw,in}$) | °C | 50 (40–70 °C change range) |
| Airflow rate ($\dot{m}_a$) | kg/h | 27.5 |
| Inlet air temperature ($T_{a,in}$) | °C | 30 |
| Inlet air humidity ratio ($\omega_{a,in}$) | g/kg | 21.5 |
| Cycle time ($t_{cyc}$) | min | 10 |
| Water flow rate ($\dot{m}_w$) | L/min | 2.5–2.7 |

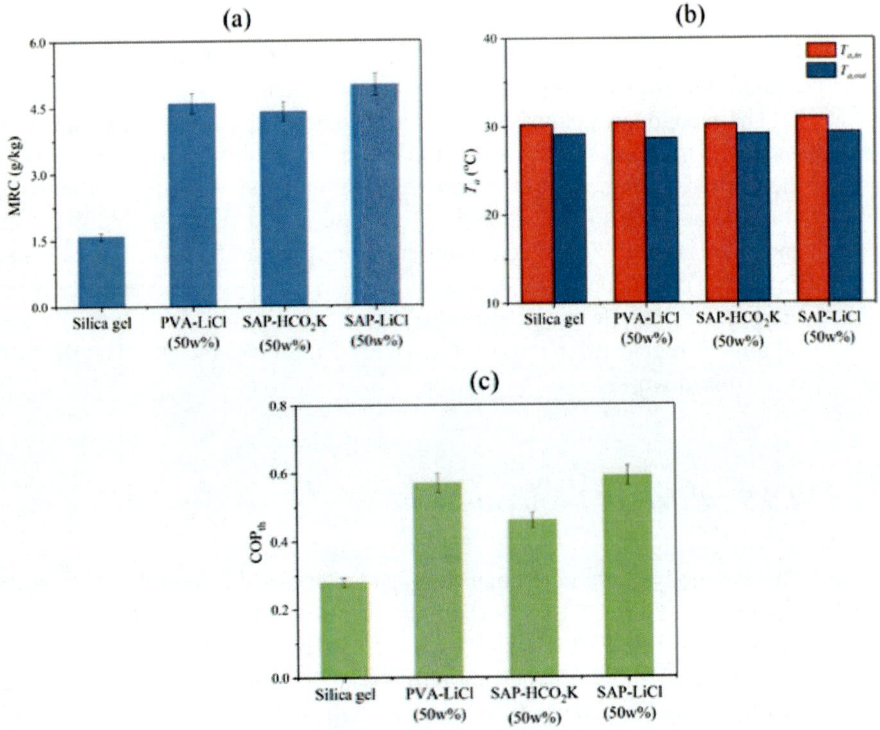


**Fig. 4.20** Effect of desiccant type on **a** MRC, **b** air temperature, and **c** $COP_{th}$, evaluated at the baseline conditions as listed in Table 4.7

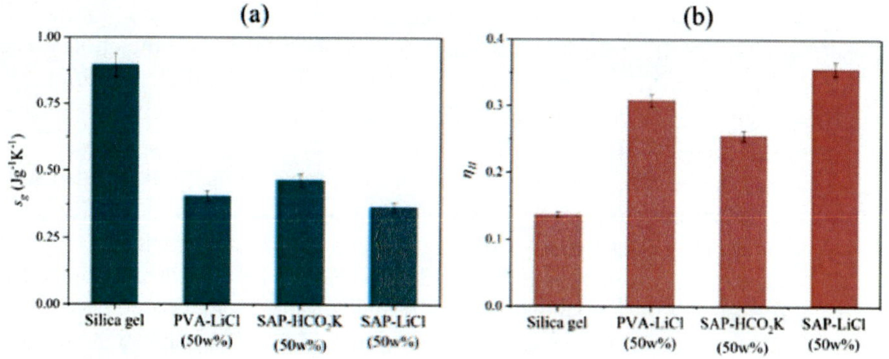


**Fig. 4.21** Effect of desiccant type on **a** $s_g$; and **b** $\eta_{II}$ evaluated at the baseline conditions as listed in Table 4.7

Figure 4.21a conveys that the silica gel-based DCHE generates 35% higher $s_g$ than the composite polymer desiccants. This is attributed to the exergy destruction magnitude during the regeneration process and the desiccant's poor MRC. Among the polymer desiccants, only SAP-$HCO_2K$ (50 w%) records increased $s_g$. This is due to lower dehumidification capacity as compared to PVA-LiCl (50 w%) and SAP-LiCl (50 w%) coated heat exchangers. The second law efficiency results are shown in Fig. 4.21b. SAP-LiCl (50 w%) records the highest $\eta_{II}$, about 2.6 times higher than silica gel. It is worthy to note that a commercial MVC air-conditioner with a COP = 3 and operating between the same thermodynamic states as SAP-LiCl (50 w%) would record about 7% $\eta_{II}$ and would need around 6 times higher amount of work input.

## 4.6 Conclusions

In this chapter, advanced engineering concepts of transport phenomena and the first and second laws of thermodynamics are applied to evaluate the performance of DCHEs theoretically. Firstly, a transient 3D mathematical model is developed by employing the fundamental conservation laws of mass, momentum, energy, and species. The normalized root mean squared error between simulated results and experimental data ranges between 7 and 12% for SGCHEs and PLCHEs. This indicates the model's excellent adaptability to predict the performance of DCHEs. The two-tube and annular fin configurations with 7 and 40% improved dehumidification performance are more effective than the existing single fin-tube configuration.

Further, the first law of thermodynamics is coupled with economic analysis to compute the energy and cost benefits realized by DCHEs. The effect of different climatic conditions, employing both 100% fresh air and 70% recirculated air configurations, multiple DCHE stages, and cooling towers are comprehensively studied. The first law analysis shows that DCHEs cut the electrical energy consumption by

40–50% under 100% fresh air configuration. When 70% of the room return is recirculated, DCHEs trim 20–30% additional cooling load from the chiller. The yearly savings accumulated by employing a DCHE in a central air-conditioning system spans 40–65% returns on the $CAPEX_{surplus}$. Accordingly, a shorter payback period of less than 2 years can be attained.

Lastly, the experimental results of DCHEs are judiciously analyzed via the second law of thermodynamics approach. The second law analysis shows that the regeneration process contributes to the highest entropy production in DCHEs. However, selecting a low regeneration temperature affects the desiccant's ability to capture moisture. Therefore, maintaining the regeneration temperature around 40 °C is considered favorable. The composite polymer DCHEs achieve 2–3 times higher first and second law efficiency than silica gel coated heat exchangers. Besides, they require 6 times lower work input as compared to the MVC air-conditioners operating between identical thermodynamic states.

## References

1. Vivekh P, Bui DT, Kumja M, Islam MR, Chua KJ. Theoretical performance analysis of silica gel and composite polymer desiccant coated heat exchangers based on a CFD approach. Energy Convers Manag. 2019;187:423–46. https://doi.org/10.1016/j.enconman.2019.02.093.
2. Ge TS, Dai YJ, Wang RZ, Peng ZZ. Experimental comparison and analysis on silica gel and polymer coated fin-tube heat exchangers. Energy. 2010;35:2893–900. https://doi.org/10.1016/j.energy.2010.03.020.
3. Oh SJ, Ng KC, Chun W, Chua KJE. Evaluation of a dehumidifier with adsorbent coated heat exchangers for tropical climate operations. Energy. 2017. https://doi.org/10.1016/j.energy.2017.02.169.
4. Jagirdar M, Lee PS. Mathematical modeling and performance evaluation of a desiccant coated fin-tube heat exchanger. Appl Energy. 2018;212:401–15. https://doi.org/10.1016/j.apenergy.2017.12.038.
5. Ng KC, Chua HT, Chung CY, Loke CH, Kashiwagi T, Akisawa A, et al. Experimental investigation of the silica gel-water adsorption isotherm characteristics. Appl Therm Eng. 2001;21:1631–42. https://doi.org/10.1016/S1359-4311(01)00039-4.
6. Chua HT, Ng KC, Chakraborty A, Oo NM, Othman MA. Adsorption Characteristics of Silica Gel+ Water Systems. J Chem Eng Data. 2002;47:1177–81. https://doi.org/10.1021/je0255067.
7. Xie X, Li D, Tsai T-H, Liu J, Braun P V, Cahill DG. Thermal conductivity, heat capacity, and elastic constants of water-soluble polymers and polymer blends;2016. https://doi.org/10.1021/acs.macromol.5b02477.
8. Zografos AI, Martin WA, Sunderland JE. Fluid properties as a function of temperature. Comput Methods Appl Mech Enginering 1987;61. https://doi.org/10.1016/0045-7825(87)90003-X.
9. Thommes M, Kaneko K, Neimark AV, Olivier JP, Rodriguez-Reinoso F, Rouquerol J, et al. Physisorption of gases, with special reference to the evaluation of surface area and pore size distribution (IUPAC Technical Report). Pure Appl Chem. 2015;87:1051–69. https://doi.org/10.1515/pac-2014-1117.
10. Vivekh P, Bui DT, Kumja M, Islam MR, Chua KJ. Theoretical performance analysis of silica gel and composite polymer desiccant coated heat exchangers based on a CFD approach. Energy Convers Manag. 2019;187. https://doi.org/10.1016/j.enconman.2019.02.093.
11. Sircar S, Hufton JR. Why does the linear driving force model for adsorption kinetics work? Adsorption. 2000;6:137–47. https://doi.org/10.1023/A:1008965317983.

12. Flossmann AI, Hall WD, Pruppacher HR. A theoretical study of the wet removal of atmospheric pollutants. Part I: the redistribution of aerosol particles captured through nucleation and impaction scavenging by growing cloud drops. J Atmos Sci. 1985. https://doi.org/10.1175/1520-0469(1985)042<0583:ATSOTW>2.0.CO;2.
13. COMSOL Multiphysics® v. 5.3 2014.
14. Vivekh P, Islam MR, Chua KJ. Experimental performance evaluation of a composite superabsorbent polymer coated heat exchanger based air dehumidification system. Appl Energy 2020; 260:114256. https://doi.org/10.1016/j.apenergy.2019.114256.
15. Ali Mandegari M, Pahlavanzadeh H, Farzad S. Energy approach analysis of desiccant wheel operation. Energy Syst. 2014. https://doi.org/10.1007/s12667-013-0115-z.
16. Shirey D, Henderson H, Raustad R. Understanding the dehumidification performance of air-conditioning equipment at part-load conditions, vol. 0641; 2006.
17. Code on periodic energy audit of building cooling system, Singapore; 2016.
18. Kalogirou SA. Solar energy engineering: processes and systems, 2nd ed. Elsevier Inc.; 2014. https://doi.org/10.1016/C2011-0-07038-2.
19. Stephens B, Novoselac A, Siegel JA. The effects of filtration on pressure drop and energy consumption in residential HVAC systems (rp-1299). HVAC R Res. 2010;16:273–94. https://doi.org/10.1080/10789669.2010.10390905.
20. Marletta L. Air conditioning systems from a 2nd law perspective. Entropy. 2013;15:859–77. https://doi.org/10.3390/e12040859.
21. Cui X, Islam MR, Chua KJ. An experimental and analytical study of a hybrid air-conditioning system in buildings residing in tropics. Energy Build. 2019;201:216–26. https://doi.org/10.1016/j.enbuild.2019.06.028.
22. Cui X, Mohan B, Islam MR, Chua KJ. Investigating the energy performance of an air treatment incorporated cooling system for hot and humid climate. Energy Build. 2017;151:217–27. https://doi.org/10.1016/j.enbuild.2017.06.059.
23. Vivekh P, Bui DT, Islam MR, Zaw K, Chua KJ. Experimental performance and energy efficiency investigation of composite superabsorbent polymer and potassium formate coated heat exchangers. Appl Energy 2020;275:115428. https://doi.org/10.1016/j.apenergy.2020.115428.
24. Vivekh P, Bui DT, Islam MR, Zaw K, Chua KJ. Experimental performance and energy efficiency investigation of composite superabsorbent polymer and potassium formate coated heat exchangers. Appl Energy 2020;275. https://doi.org/10.1016/j.apenergy.2020.115428.
25. Sodium Polyacrylate cross-linked|2-propenoic acid, homopolymer, sodium salt|Sigma-Aldrich n.d. https://www.sigmaaldrich.com/catalog/product/aldrich/432784?lang=en&region=SG&gclid=Cj0KCQiA4NTxBRDxARIsAHyp6gCIrpBXD9YgqzB7sNgXKq8eYJEI8W5h6k9wktL8TtBB0R_KVw7qY_8aAiLtEALw_wcB. Accessed 7 Feb 2020.
26. Potassium formate|Sigma-Aldrich. n.d. https://www.sigmaaldrich.com/catalog/substance/potassiumformate8412590294111?lang=en&region=SG. Accessed 7 Feb 2020.
27. Electricity Tariffs|EMA Singapore. n.d. https://www.ema.gov.sg/Residential_Electricity_Tariffs.aspx. Accessed 7 Feb 2020.
28. Power Consumption of Typical Household Appliances. n.d. https://www.daftlogic.com/information-appliance-power-consumption.htm. Accessed 7 Feb 2020.
29. Cooling Tower Replacement Cost—All Kote Lining, Inc. n.d. https://allkoteliningincaz.com/cooling-tower-replacement-cost/. Accessed 7 Feb 2020.
30. Vivekh P, Bui DT, Islam MR, Zaw K, Chua KJ. Experimental performance evaluation of desiccant coated heat exchangers from a combined first and second law of thermodynamics perspective. Energy Convers Manag. 2020. https://doi.org/10.1016/j.enconman.2020.112518.
31. Chengqin R, Nianping L, Guangfa T. Principles of exergy analysis in HVAC and evaluation of evaporative cooling schemes. Build Environ. 2002;37:1045–55. https://doi.org/10.1016/S0360-1323(01)00104-4.
32. Bejan A. Adv Eng Thermodyn. 2016. https://doi.org/10.1002/9781119245964.
33. Bejan A. Second law analysis in heat transfer. Energy. 1980. https://doi.org/10.1016/0360-5442(80)90091-2.

34. Alhazmy MM. The minimum work required for air conditioning process. Energy. 2006;31:2403–13. https://doi.org/10.1016/j.energy.2005.12.007.
35. Saul A, Wagner W. International equations for the saturation properties of ordinary water substance. J Phys Chem Ref Data. 1987. https://doi.org/10.1063/1.555787.
36. Stoecker WF (Wilbert F., Jones JW (Jerold W.)). Refrigeration and air conditioning. McGraw-Hill; 1982.
37. Lior N, Al-Sharqawi HS. Exergy analysis of flow dehumidification by solid desiccants. Energy. 2005;30:915–31. https://doi.org/10.1016/j.energy.2004.04.029.
38. Vivekh P, Bui DT, Wong Y, Kumja M, Chua KJ. Performance evaluation of PVA-LiCl coated heat exchangers for next-generation of energy-efficient dehumidification. Appl Energy. 2019;237:733–50. https://doi.org/10.1016/j.apenergy.2019.01.018.

# Chapter 5
# Advanced Energy-Related Applications of Desiccants

**Abstract** The enhanced energy performance of desiccant dehumidifiers through new material synthesis and optimized design have extended their potential to several state-of-the-art energy-related applications including heat transformation, adsorption chilling, energy storage, and water harvesting. A review of the latest research trends in these system-level applications is presented in this chapter. Additionally, controlling methods to yield precise moisture levels for human thermal comfort and various industrial manufacturing processes are described. For human well-being, both extreme low and high humidity levels can cause health issues such as dryness and nausea including critical illnesses due to the airborne spread of bacteria, fungal, and viral infections. In industrial operations, the presence of high moisture levels may damage machines, deteriorate product quality, and ultimately lead to significant revenue losses. For such critical applications, desiccant dehumidifiers offer close and precise moisture control in an energy-efficient and cost-effective manner.

**Keywords** Heat pumps · Adsorption chillers · Industrial dehumidifiers · Atmospheric water harvesting · Energy storage · Indoor moisture control

## 5.1 Introduction

Essential investigations on both experimental and theoretical performance analyses summarize the key improvements realized by adopting superior materials and more effective parameter selection. Advanced engineering approaches further facilitate improved understanding on the suitability of coupling desiccant dehumidifiers with numerous heat transfer/exchange devices. As a result, desiccant dehumidifiers are ready for immediate implementation in wide-ranging energy-related applications, specifically classified into four categories based on their key objectives, as summarized in Fig. 5.1. This chapter presents the potential applications of desiccants, analyzes its relevance, and highlights the latest research directions in these fields. The existing building and industrial air-conditioning case studies are then presented, and the factors required for designing a desiccant-enhanced cooling system are also specified.



V. Prabakaran and K. J. Chua, *Advances in Desiccant Dehumidification*,
https://doi.org/10.1007/978-3-030-80843-3_5

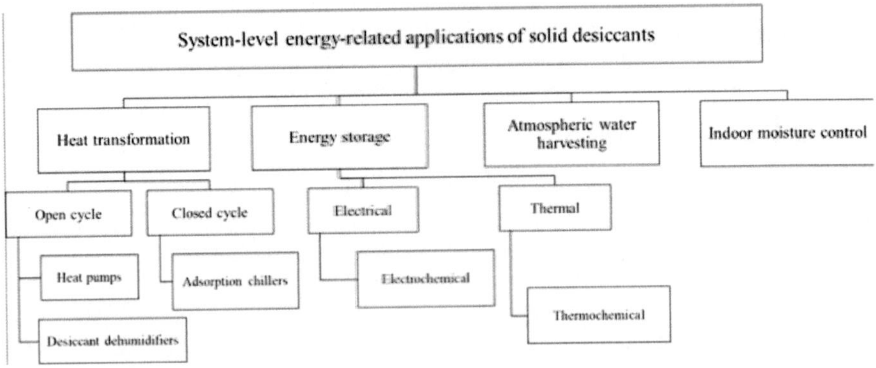


**Fig. 5.1** Classification of the system-level applications of solid desiccants

## 5.2 An Overview of Potential Applications

The energy-related applications of solid desiccants can be grouped into four domains: heat transformation, energy storage, water harvesting, and humidity control. In this section, a comprehensive review is first presented for these applications by establishing their relevance to energy efficiency and environmental issues.

### *5.2.1 Heat Transformation Applications*

The heat transformation applications are classified as either open or closed-cycle systems depending on whether the working fluid operates in open/closed cycles. Open cycle systems are directly coupled with the ambient air and operate in an indefinite loop. Standalone desiccant dehumidifiers/hybrid desiccant air-conditioning systems, and desiccant enhanced heat pumps are technologies that comprise the open-cycle systems. In contrast, closed-cycle systems include adsorption chillers. The working fluid operates in a closed cycle by undergoing evaporation, condensation, adsorption, and desorption processes. The standalone desiccant dehumidifiers have already been discussed earlier in Chap. 1, therefore, the ensuing section on open-cycle application primarily focuses on desiccant coated heat pumps. For the closed-cycle systems, current developments in adsorption chillers are reviewed.

#### 5.2.1.1 Desiccant-Coated Heat Pumps

Conventional direct expansion heat pumps are modified into desiccant-coated heat pumps (DCHPs) by replacing the evaporators and condensers with two identical DCHEs, which function as desiccant-coated evaporators and condensers interchangeably. A schematic representation of the working principle of the DCHPs is provided in

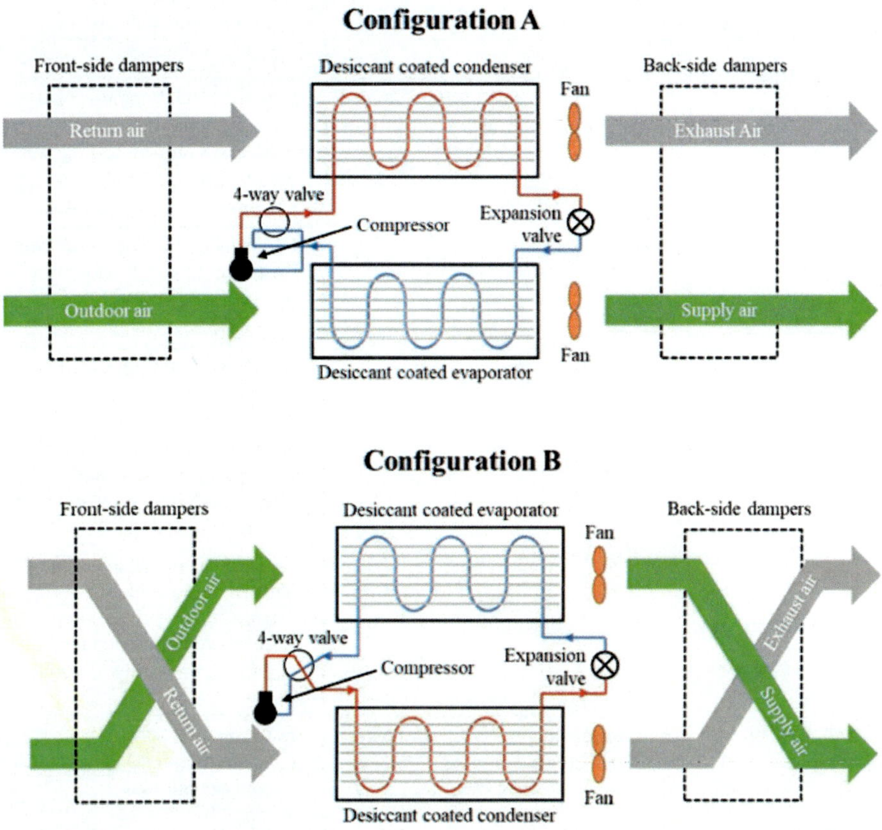


**Fig. 5.2** A schematic exhibiting the working principle of a desiccant coated heat pump system [12]

Fig. 5.2. The system is charged with a chemical refrigerant (with low GWP selection criteria) that undergoes the vapor-compression refrigeration cycle and is controlled through the incorporation of a four-way valve. The outdoor air passes through the desiccant-coated evaporator and is delivered to the desired cooling space. The return air is supplied to the desiccant-coated condenser to regenerate the DCHE and then exhausted. The respective process and the exhaust air are supplied to the evaporator and the condenser using the front-side air dampers. The air outlet from the evaporator and condenser are directed appropriately by controlling the back duct air dampers. A list of the different types of desiccants used in DCHPs, the type of study conducted, their key attributes are summarized in Table 5.1, and Fig. 5.3 shows a photograph and schematic of MOF-based DCHPs.

**Table 5.1** A review of different desiccants used for heat pumping applications

| Literature | Desiccant type | Type of study and key results |
|---|---|---|
| [1] | Composite silica gel (16.2w%) with emulsion glue binder | • Experimental dynamic performance evaluation was conducted<br>• A loosely coupled heat and mass transfer approach model was developed<br>• System COP of 6.2 could be achieved |
| [2] | Silica gel | • Experimental studies were carried out in various outdoor air and return air conditions of the Asia-Pacific Region<br>• A COP of 5.7 was achieved |
| [3] | AQSOA Z02 zeolite | • A DCHP system for electric vehicles was proposed<br>• Simulation of heating mode at low ambient temperatures was performed<br>• De-fogging of the windshield under interior-air circulation and a largest cruising range could be achieved |
| [4, 5] | Silica gel | • Both experimental and theoretical performance evaluation were conducted<br>• The novel system can supply air with lower temperature and temperature variation comparing to traditional heat pump |
| [6–8] | Composite silica gel LiCl (38w%) with emulsion glue binder | • Experimental performance studies were conducted under ARI and Shanghai summer conditions<br>• A three-dimensional mathematical model was developed<br>• Maximum system COP of 7.0 can be reached |
| [9] | Aluminum fumarate (Basolite A520) MOF with polysiloxane binder | • Experimental study on isothermal capacity was conducted<br>• 20–30% improvement in adsorption kinetics in contrast to SAPO-34 coated heat exchanger<br>• A cooling energy efficiency of around 60% could be achieved |
| [10] | MIL-100(Fe) with a warne-borne silica sol binder | • The adsorption heat is comparable to the latent heat of vaporization of water<br>• A temperature difference of around 30 °C between the evaporator and the condenser was sufficient to exploit the S-shape isotherm of MOFs<br>• A system COP of 7.9 could be attained when the condensation temperature dropped to 45 °C |

(continued)

**Table 5.1** (continued)

| Literature | Desiccant type | Type of study and key results |
|---|---|---|
| [11] | Composite superabsorbent polymer desiccant (50w% LiCl) | • Experimental study on performance evaluation and a simplified one-dimensional lumped parameter model for performance evaluation<br>• The mathematical model displayed excellent prediction accuracy and was independent to the type of desiccant material, airflow configuration, and nature of refrigerant<br>• DCHPs yielded 30–40% higher COP than conventional heat pumps |

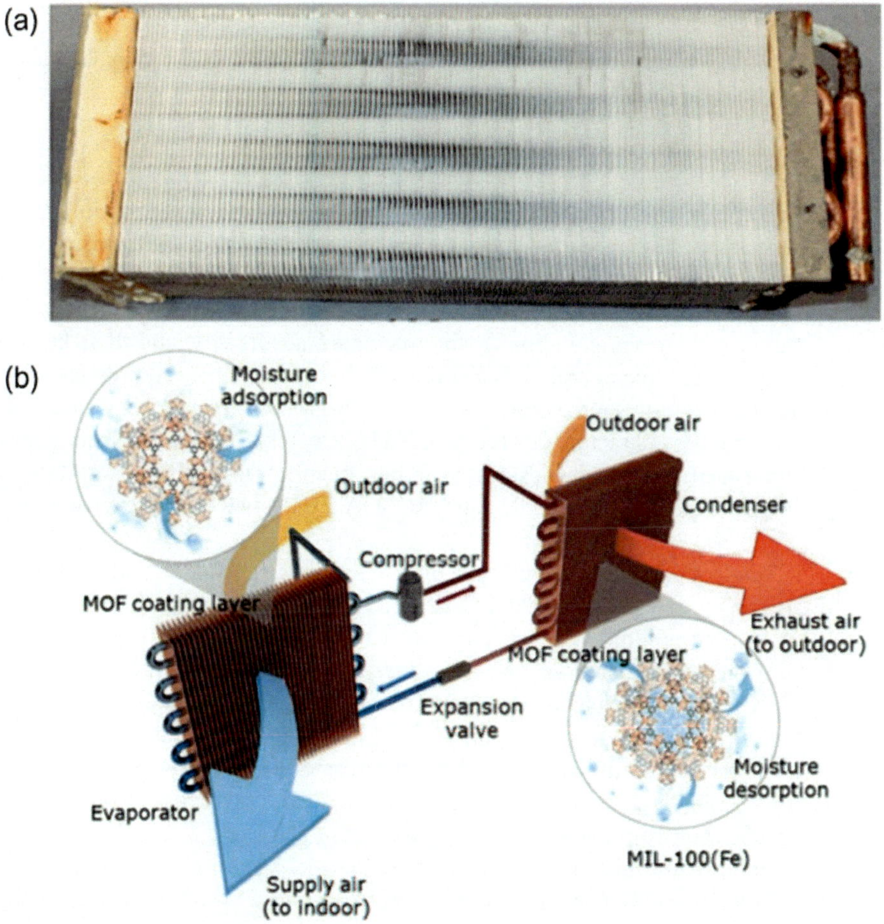


**Fig. 5.3** A photograph of DCHEs based on **a** aluminum fumarate [9] and **b** schematic representation of MIL-100(Fe)-based cooling system [10]

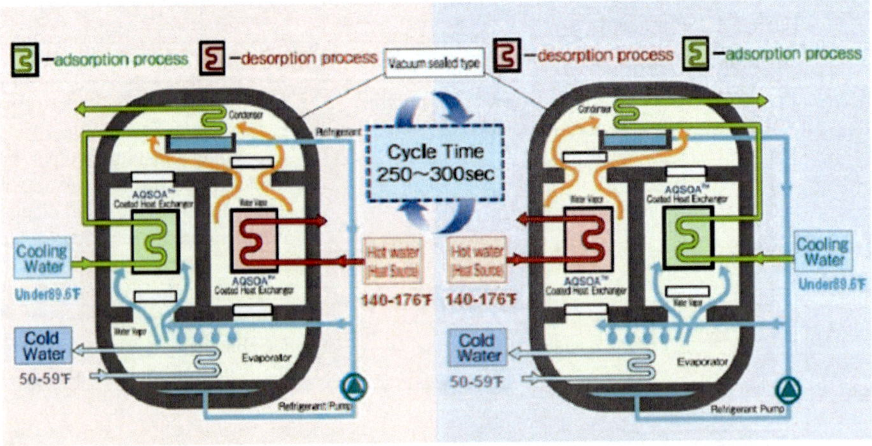


**Fig. 5.4** A schematic of AQSOA zeolite coated heat exchanger-based adsorption chiller [15]

#### 5.2.1.2 Adsorption Chillers

Adsorption chillers operate on the reversible adsorption-desorption principle, as previously described in Chap. 1. Several adsorption chillers based on packed-bed dehumidifiers have been developed using silica gel, composite silica gel, zeolites, alumina, and activated carbon desiccants [13]. These adsorbents have been paired with water/ammonia/ethanol/methanol adsorbates, and their key disadvantages emerge due to high pressure drop, bulkiness, and the requirement of regeneration temperature sources above 80 °C [14]. Significant benefits in terms of improved heat transfer characteristics are realized by replacing the packed-bed dehumidifiers with DCHEs. Industrial adsorption chillers have emerged employing AQSOA zeolite (AQSOA-Z01, AQSOA-Z02, and AQSOA-Z05) coated heat exchangers [15, 16] (Fig. 5.4).

Improved isothermal capacity compared to silica gel and high degree of durability are the key advantages of AQSOA coated heat exchangers [17], while high cost and small pore volume are their competitive drawbacks [18]. Z01 was established to be a better candidate against Z02 since its regeneration temperature was around 65 °C and Z02 demanded temperatures over 80 °C [19]. In addition, based on the heat of adsorption measurements, Z05 showed the lowest adsorption heat implying its higher suitability [20]. Recently, the feasibility of using MOFs for adsorption chilling applications has been rigorously studied as the cost of producing them in bulk is sharply reducing. MIL-53(Al)-TDC (TDC = 2,5-thiophenedicarboxylate) exhibited a low adsorption heat value (only 2.4 kJ/kg). In addition, Fig. 5.5 illustrates that the cycle could operate at a temperature swing of 15–55 °C, achieving maximum water uptake capacity at 15 °C and adequate desorption at a moderate regeneration temperature. These factors lead to an easier integration with waste heat sources, which are employed to thermally power the adsorption chillers [21].

MIL-101(Cr) and silica composite material were used for industrial adsorption chilling applications, and it exhibited excellent water adsorption results, albeit for

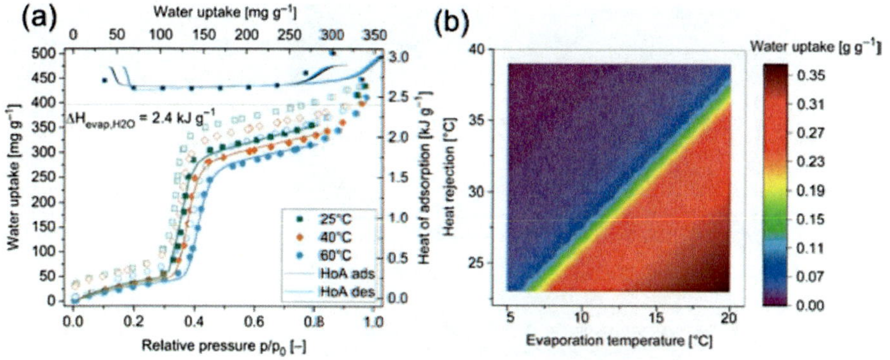


**Fig. 5.5** **a** Water sorption isotherms at three different temperatures, and the heat of adsorption in MIL-53(Al)-TDC; **b** water uptake capacity in MIL-53(Al)-TDC as a function of waste heat and evaporation temperature [21]

merely 8 cycles [22]. In another study, several Zr-based MOFs like UiO-66, UiO-67, and NU-1000 were adopted for adsorption chillers, and their performance was analyzed via experimental and modeling approaches [23]. Water and ethanol working pairs were chosen for the analysis. Water demonstrated higher compatibility for smaller pore-sized MOFs due to its high specific cooling capacity and COP at a lower temperature. In contrast, ethanol remained the preferred choice for MOFs with a larger pore size due to high uptake capacity and cyclic stability. Further, an experimental investigation of another MOF (CAU-10(H)) in adsorption chillers was carried out [24]. Monoliths of various shapes were prepared (Fig. 5.6a), and excellent cyclic stability was obtained with no significant performance variation for more than 10,000 adsorption-desorption cycles. A water-based SilRes MP50E binder was employed, and the changes in the water adsorption capacity remained inconsequential. Also, a mild regeneration temperature of 70 °C was employed, pointing towards straightforward implementation with low-grade waste heat or solar thermal collectors.

### 5.2.2 *Energy Storage*

A reliable energy storage system is necessary to attain the economic feasibility of renewable energy sources because there is a mismatch between the peak electricity/cooling requirement and energy availability. Electrochemical storage systems store energy as a chemical potential of the material and discharge electricity, are widely adopted in electric vehicles and mobile phones and achieve localized zero-energy emissions. Certain MOFs have been studied as alternative materials in various electrochemical storage technologies such as lithium-ion, sodium-ion batteries, electrochemical supercapacitors, and fuel cells [26]. A comprehensive list of different MOFs and their structural characteristics can be found in the literature [27–29].

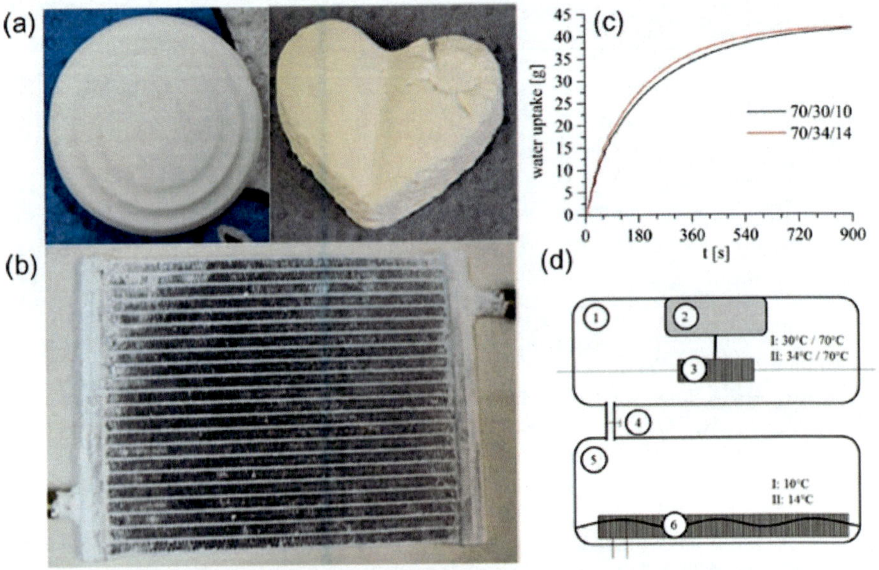


**Fig. 5.6** **a** Different shapes of monoliths prepared from MOF/binder composites, **b** photograph of a coated heat exchanger, **c** water uptake of a coated heat exchanger, **d** prototype measurement device with adsorption chamber, balance, adsorber with hydraulic connections [24]

Thermal energy storage systems can be classified into sensible, latent, and thermochemical systems. Materials with a high specific heat capacity and thermal conductivity are preferred in sensible heat storage systems. Water/brine solutions, often used in solar thermal systems, are among the most widely used fluids in sensible heat storage systems. A substantially higher amount of energy can be stored as latent heat by utilizing phase-changing properties of specialized materials. Although ice slurry and paraffin waxes are naturally occurring materials for latent heat storage, organic/inorganic materials are also being synthesized. Despite the fact that sensible and latent heat storage systems are widely employed, they possess low storage density, i.e., less than 50 kWh/m$^3$ for sensible [30] and less than 100 kWh/m$^3$ for latent heat storage systems [31]. Moreover, temperature restrictions and heat losses are their major drawbacks. Since a perfect insulation is often difficult to achieve, the inevitable heat loss markedly lowers the system's performance. Thermochemical energy storage systems, on the other hand, store heat in the form of the employed material's chemical potential. These systems need substantially less space because their energy density ranges spanning 200–500 kWh/m$^3$ [31], and the energy storage occurs via the reversible physisorption processes.

A porous adsorbent (with a large specific area) and a fluid adsorbate (with a high latent heat of evaporation) constitute the thermochemical energy storage device. It comprises three phases of operation: desorption, storage, and adsorption. Since desorption is an endothermic process, the adsorbent stores the heat while releasing the adsorbate. This is referred to as the discharged state of the adsorbent. Upon

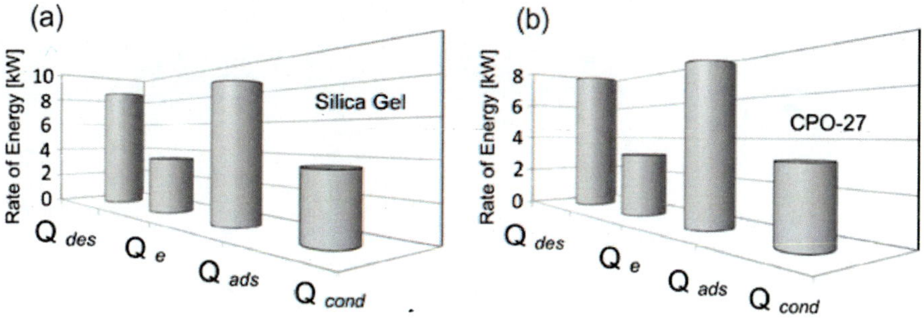


**Fig. 5.7** Energy distribution during cooling mode in **a** silica gels and **b** CPO-27(Ni) [37] where $Q_{des}$ is the rate of heat absorbed by the desiccant during desorption; $Q_e$ is the energy rate removed by the evaporator; $Q_{ads}$ is the rate of heat released by the desiccant during adsorption; and $Q_{cond}$ is the heat rate rejected by the condenser

adsorption, the pores of the adsorbent are filled with the adsorbate molecules. After establishing an equilibrium with its surroundings, the adsorbent continues to remain in the charged state indefinitely. No thermal losses occur during the storage step since the desorption and adsorption steps can be easily decoupled [32]. When using water as adsorbate, the thermochemical energy storage systems may be connected to either adsorption chillers or heat pumps. Conventional solid desiccants such as silica gel, hygroscopic salt-infused silica gel [33], zeolite 5A [34], and zeolite 13X [35] have been widely used, whereas the feasibility of employing advanced materials like MOFs are not yet comprehensively understood [36]. Due to the ease of large-scale production, Elsayed and co-workers [37] used CPO-27 (Ni) MOF for energy storage. Its energy density was measured to be 170.2 Wh/kg at a charging temperature of 100 °C, around three times of zeolite 5A. Although both silica gel regular density (RD) powder and CPO-27(Ni) have a comparable energy density, CPO-27(Ni) adsorbs 80% more moisture than silica gel RD at lower evaporator temperatures, making it a feasible candidate (Fig. 5.7). Further, the impregnation of composite salts in MOFs such as $CaCl_2$ in UiO-66 MOF demonstrated a specific cooling power of 0.63 kW/kg and a COP of 0.83 [38]. A record-high isothermal water uptake capacity of around 1.93 g/g was also obtained at 90% RH, and its heat storage capacity was found to be 367 kJ/kg, a key criterion for usage in thermal batteries [39].

### 5.2.3 *Atmospheric Water Harvesting*

In arid climates, extracting potable water from the ambient air is a valuable proposition since freshwater availability is severely limited, employing desalination technologies remains uneconomical, and the potential for integrating renewable energy sources is abundant. The water harvesting from the atmosphere is carried out in a two-stage process (Fig. 5.8). In the first stage, the adsorbent captures moisture

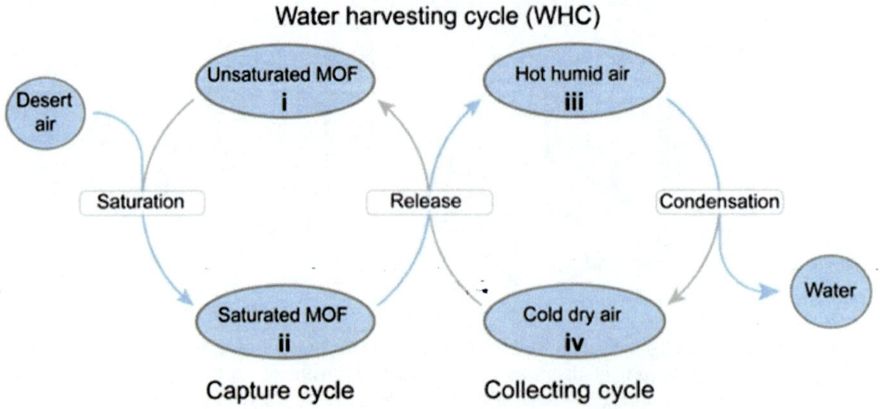


**Fig. 5.8** A schematic demonstrating the working principle of an atmospheric water harvesting system. Process i to ii indicates the reversible adsorption and desorption cycles in the desiccant based on the atmospheric condition. Process ii to iii represents the humidification of the dry air as the desiccant reactivates for the next cycle. Finally, Process iii to iv signifies the condensation process yielding liquid potable water [40]

from the ambient air (during the night) and then releases the adsorbed water vapor to the atmospheric air (during the day). Subsequently, the released water vapor is cooled to below the air dew point in a condenser through natural cooling or renewable energy/electricity-driven devices. While the proposed system seems to demonstrate excellent commercial development at first glance, efficient ways of generating water from the atmosphere have not been developed yet. The two main challenges that prevent the building of a practical system are the absence of materials that can effectively absorb and release a large amount of water vapor and the need for energy-intensive sources for condensing the absorbed water vapor.

The desiccant's sorption-desorption characteristics play a key role in the device's success. Conventional desiccants, including silica gel and zeolites for water harvesting, are not feasible due to their low adsorption ability and high regeneration energy consumption. The desired characteristics for water harvesting include a steep rise in water uptake within a narrow RH range, high cyclic stability, tailorable hydrophilicity, and excellent sorption kinetics [41]. Yaghi and his co-workers [42] investigated the ability of MOF-801 $\{Zr_6O_4(OH)_4(fumarate)_6\}$ coated copper foam to extract water from the atmosphere at low RH climatic conditions. A non-concentrated solar heat flux (with solar intensity less than 1 $kW/m^2$) was used for releasing the water vapor captured by MOF. The released water vapor was then condensed to attain potable water via a condenser maintained at ambient temperature. Figure 5.9a illustrates their prototype and shows the transient water droplet formation at different temperatures. Although this proof-of-concept study exhibited excellent theoretical potential, Bui and co-workers [43] argued that the prototype incurred high heat losses (90–94% of the solar energy dissipation) due to its poor

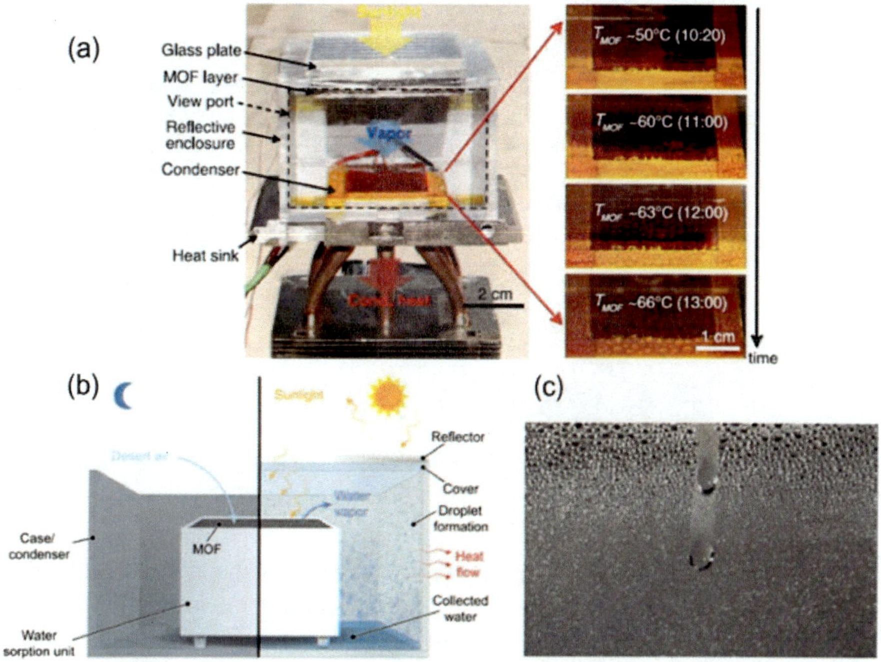


**Fig. 5.9** **a** A picture of the water harvesting device developed by Kim and co-researchers [42] and water droplet formation at different time and temperature; **b** A schematic of the device developed for water harvesting by Fathieh and co-workers [40]; **c** A picture of the actual water droplet formation

design. Yaghi and his co-researchers [44] concurred that to make the prototype technologically feasible and economically viable, numerous optimization strategies must be carried out and its design must be improved.

To increase the water yield produced by the prototype in [42], Fathieh and co-researchers [40] used 1.2 kg of MOF-801 and fabricated a water harvesting system (Fig. 5.9b). They performed laboratory-scale experiments and real-time tests in the Arizona deserts. Sunlight and natural cooling were the employed energy sources. MOF-801 was completely saturated at 30–50% RH and 18–25 °C at night in laboratory conditions. However, the actual desert conditions were 5% RH at 35–40 °C during the day and reached up to up to 40% RH at 10–15 °C during the night. The system extracted water with a 33% overall efficiency in laboratory conditions. The overall efficiency was defined as the ratio of the mass of potable water collected to the amount of water vapor captured by MOFs. To collect water with a similar efficiency under the actual desert environment, the system had to be modified by adding an insulation layer and tilting the device's angle to 37° to absorb maximum solar irradiance.

A novel aluminum-based MOF-303 desiccant was also synthesized in their study, and it yielded around three times higher than MOF-801 and zeolite 13X. The MOF-303-based device generated 1.3 and 0.7 L of water per kg of MOF used under laboratory and actual desert conditions. Through these experiments, it was established that

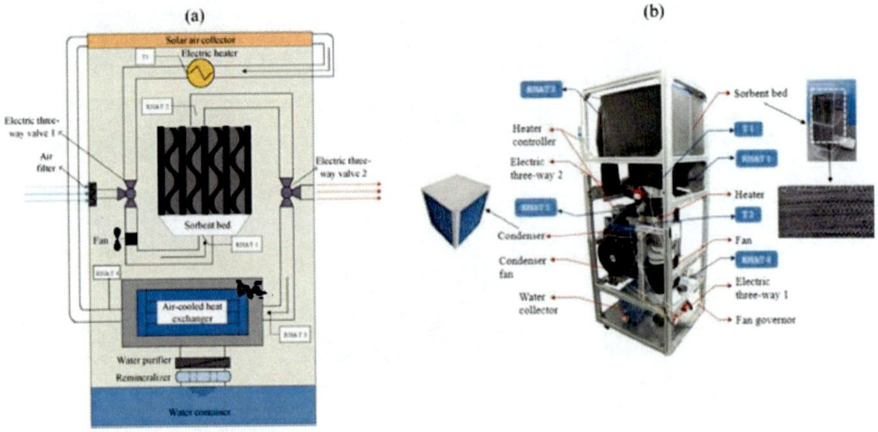


**Fig. 5.10** **a** A schematic and **b** photograph of the device for harvesting atmospheric water in an island [45]

synthesizing sorbents with faster kinetics remained the key to achieve atmospheric water harvesting matching the scale of human water consumption [25]. In another study, to propose a technologically viable water harvesting solution for an island (maintained at 31 °C and 63% RH), Wang and co-workers [45] built a device using activated carbon fibers and silica sol-LiCl sorbent (Fig. 5.10). By employing an air-cooled condenser and no additional cooling source, this device produced 7.7 kg of potable water when 21 kg of the desiccant was used.

### 5.2.4 *Moisture Control*

Achieving precise humidity control is a key consideration to thermal comfort of occupants in buildings and conducting specific industrial/manufacturing tasks. While some industrial operations need ultra-low humidity levels (discussed in Sect. 5.4), the humidity set point for thermal comfort in a standard building environment ranges spanning 40–60% RH [46]. High humidity promotes the growth of molds, bacteria, and dust mites and leads to several respiratory illnesses. While these factors contribute to degrading occupants' health, excess moisture levels also induce undesired chemical reactions that affect building materials and wooden structures/furniture [47]. Solid desiccants have been used for indoor moisture management due to the desiccant's high water sorption ability and ease of regeneration at low temperatures. Desiccant dehumidifiers can be employed to provide localized humidity control, retain appropriate moisture at indoor humidity levels, and be functional in enclosed spaces, such as aircraft and submarines. While silica gel, zeolites, and activated carbons have been used to provide localized moisture control [48], MOFs have also turned out to be attractive. Their applicability has primarily been due to the high-water uptake, S-shaped isotherms, and increased stability of around 1000 cycles.

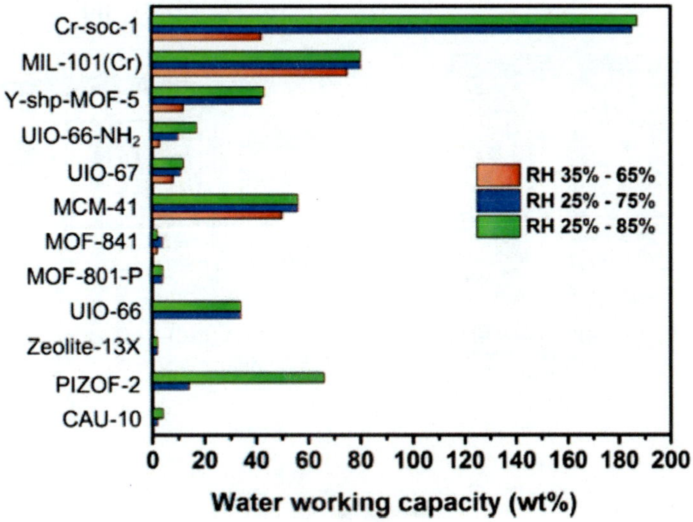


**Fig. 5.11** Water uptake capacity of different MOFs at different RH levels, typically used in humidity control applications [49]

Figure 5.11 documents a list of MOFs used in humidity control applications and highlights their water uptake capacities at multiple humidity levels [49].

## 5.3 Thermal Comfort: Case Studies on Humidity Control

A review of advanced applications of the desiccant dehumidifiers has been presented in the preceding sections. In this section, the use of desiccant dehumidifiers is presented for different cases of thermal comfort, including indoor and outdoor environments under tropical climatic conditions. Numerous sizes of rotary wheel dehumidifiers are available for specific applications. Their size corresponds to the amount of air handled, typically ranging from 200 $m^3$/h to as high as 20,000 $m^3$/h, as portrayed in Fig. 5.12.

### *5.3.1 Indoor Cooling*

In transient indoor cooling applications such as classrooms, lecture theatres, and conference facilities, the cooling load changes rapidly during the day because of the frequent inflow/outflow of people. As a result, the building cooling load demand prediction is an essential step to determine the control strategy of the employed air-conditioning unit. Besides regulating temperature and humidity to desired thermal comfort standards, maintaining appropriate indoor air quality (IAQ) is also vital to

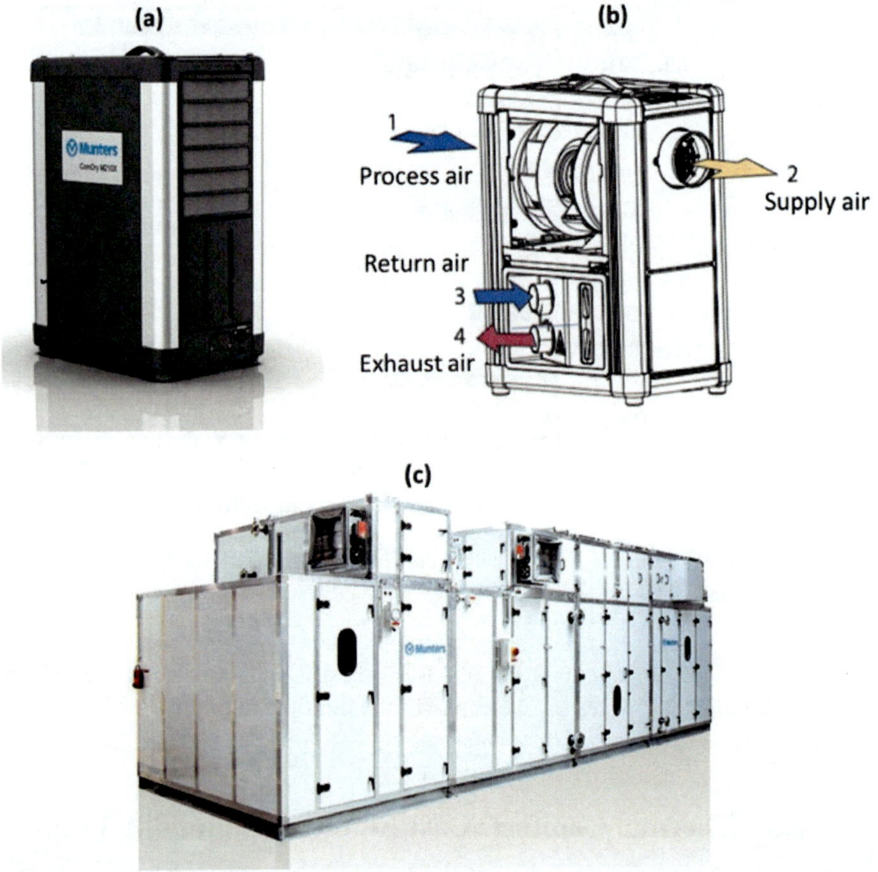


**Fig. 5.12** **a** A standalone compact desiccant dehumidifier [50]; **b** the working principle of the compact dehumidifier [50]; and **c** a large-scale desiccant dehumidifier for central air-conditioning systems [51]

prevent bacterial/fungal/viral airborne respiratory infections. The particles causing respiratory illnesses range in numerous sizes (less than 1 μm to greater than 50 μm), and those with less than 10 μm remain airborne for several hours [52]. Specifically, with the coronavirus infectious disease (COVID-19) pandemic that has infected more than 150 million people and has claimed the lives of close to 3.2 million at the time of writing this text [53], improved IAQ standards have become more indispensable.

As scientists worldwide continue to study the effect of humidity on propagating infectious diseases, ASHRAE has recommended to maintain RH between 40–60% for the maximum protection against coronaviruses (SARS-Cov-1 and SARS-Cov-2) [54]. When the indoor air RH goes below 40%, the resulting environment free of moisture improves the evaporation kinetics of the virus aerosols and provides an optimal path for long-distance transmission. In contrast, RH levels greater than 80% increase the particle size due to hygroscopic swelling and lead to higher local deposition rates [55, 56]. While MVC technology is employed for indoor air-conditioning, the cooling coil must be maintained sufficiently lower than the air's dew point temperature. So the indoor space gets excessively cooled and dry upon any drop in the cooling load demand. To address this problem, the cooling systems are switched off periodically. This negatively impacts the room IAQ and results in moldy odors. Employing a desiccant dehumidifier alongside the conventional MVC system mitigates this problem as a dedicated component of the hybrid cooling system handles the latent load requirement. The cooling coil temperature can then be increased from 6 to 7 °C to around 20 °C, thereby translating to around 30–35% rise in energy efficiency [57].

Desiccant wheel dehumidifiers have been widely deployed in hybrid cooling systems due to their superior dehumidification-to-size effectiveness compared to stationary bed dehumidifiers. At present, the desiccant wheels can either be installed upstream of the cooling coil or downstream, as demonstrated in Fig. 5.13 [54]. In the upstream configuration, the desiccant wheel dehumidifies the fresh air before passing through the cooling coil. Placing a cooling coil after the dehumidifier is necessary to remove the excess temperature due to the released sorption heat. In the downstream configuration, the outdoor air is pre-cooled via the MVC air-conditioners, and then the desiccant wheel is employed for dehumidification. The downstream arrangement ensures that the desiccant wheel captures maximum moisture from the nearly saturated air states and functions more effectively [54]. Lastly, as revealed through earlier Chaps. 1–4, the advantages of DCHEs and their easy retrofitting abilities make them highly suitable for future desiccant cooling applications. DCHEs allow both desiccant dehumidification and cooling processes to co-occur, implying that they can reduce the bulkiness of integrating two cooling technologies and potentially function as a standalone cooler.

### 5.3.2 *Outdoor Cooling*

In tropical climates, high temperature and humid outdoor weather coupled with a lack of appropriate ventilation pose significant challenges for high public density sites, such as walkways, bus stops, and alfresco dining. In these applications, the ambient air at 32–34 °C and 22–24 g/kg must be supplied at about 24–25 °C and 15–16 g/kg for adequate thermal comfort. Cooling the air to further lower temperatures does not provide significant benefits since the conditioned air invariably mixes with the ambient air and gets heated and humidified. Although many water-based direct evaporative coolers cater to this application, the air temperature is reduced at the expense of higher add-on humidity. Exposure to the humidified air may seem comfortable for a short period, but many associated health issues occur when the duration is longer. These issues include respiratory infections due to bacterial/fungal growth, skin allergies, and overheating of the human body.

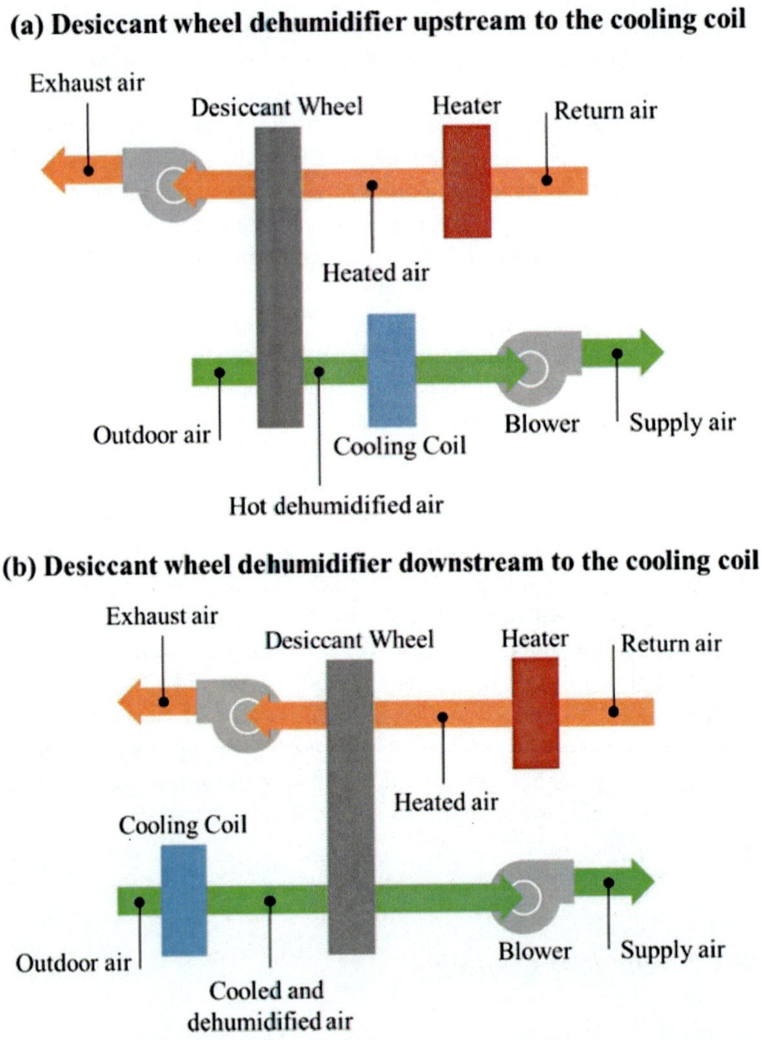


**Fig. 5.13** Desiccant wheel configurations installed **a** upstream and **b** downstream to the cooling coil [54]

MVC air-conditioners are often deemed to be not suitable for outdoor cooling since the excessive cooling of environmental air occurs. In addition, the released heat from the condenser and compressor leads to a micro-heating effect. A hybrid air cooler comprising a desiccant wheel or two DCHEs and an indirect evaporative cooler (IDEC) is appropriate for supplying both cool and dry air to an outdoor space environment. While the role of IDEC alongside desiccant wheels is more pronounced, it functions as an auxiliary cooling unit when coupled with DCHEs. The hot and humid air first passes through one of the DCHEs (operating at 20 °C), which serves as a dehumidifier. The other DCHE (at 40 °C) functions as a regenerator and releases

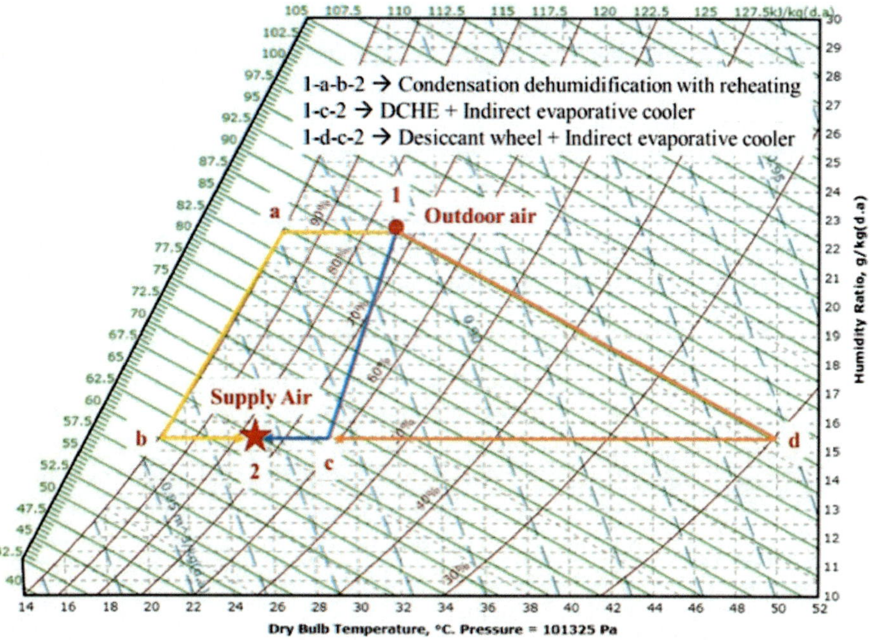


**Fig. 5.14** A psychrometric chart highlighting the three types of outdoor cooling processes. The line 1-a-b-2 indicates condensation dehumidification and cooling process followed by reheating; 1-c-2 points to DCHE dehumidification and indirect evaporative cooling; and 1-d-c-2 represents the desiccant wheel dehumidification with indirect evaporative cooling

the captured moisture. Low-temperature regeneration is favorable for DCHEs as the exhaust air closer to the ambient temperature is dissipated to the environment instead of hot air streams. The dehumidified air further passes through the IDEC, where it cools down by a water-based evaporative cooling process. Figure 5.14 illustrates a psychrometric chart with the three types of cooling processes used in outdoor cooling applications.

### *5.3.3 Healthcare Sector: Surgery/Critical Care and Inpatient Rooms*

In healthcare facilities, excellent IAQ is imperative to control infection rates and provide quick patient recovery. Since patients are exposed to a greater content of bacteria/viruses that could be airborne, higher fresh air intake or more frequent air changes are often necessary. In many hospitals, the outside air would account for 90–100% of the total cooling load, with latent load levels as high as 70% in tropical climates.

Conventional MVC cooling systems are not suitable for handling the thermal comfort requirements of critical and surgical care units. This is attributed to the fact that MVC systems adopt a condensation dehumidification approach and saturate the air. While they can be designed for the requirement of 22–24 °C in the common areas of hospitals, perioperative care units need lower temperatures in the range spanning 12–18 °C. Such low temperatures are necessary to ensure patient safety and to achieve the desired thermal comfort for surgical staff. The staff need to wear thick layers of personal protective equipment and undergo more prolonged medical procedures. As a result, lower room temperature is essential to achieve a comfortable work environment [58].

Aside from temperature, maintaining strict control of 40–50% RH levels is also critical. A low humidity level causes dryness, and humidity greater than 65–70% RH promotes the growth and multiplication of several bacteria/viruses that may cause lung infections. Further, condensate formation and water leakage promote bacterial, fungal, and viral growth in the ductwork and may trigger the spread of infection. In addition, higher moisture levels may also corrode sensitive medical equipment; leading to incorrect/false diagnoses. Achieving the right RH level at low air temperatures would require the air to be drier than the saturated air produced by the conventional MVC systems; translating to greater challenges in designing the air-conditioning system.

The fundamental characteristic of a desiccant dehumidifier allows it to reduce air humidity without moisture condensation, thus making it a viable alternative for temperature and humidity control in hospitals. The desiccant dehumidifiers can be employed to handle the latent load component of the fresh air entering the hospitals. The dehumidified air is then supplied to the existing surgical care AHU for sensible cooling. As the cooling coil only caters to sensible cooling, its temperature need not be maintained below the air's dew point to condense moisture. This makes the cooling process highly energy-efficient and dramatically reduces microbiological contamination risk [59]. Figure 5.15 illustrates the desiccant dehumidifier configurations for various facilities in the hospitals.

## 5.4 Industrial Air-Conditioning

The role of climate control is critical in order to achieve the desired return on investment and appropriate functioning of different stages of key industrial activities, namely, manufacturing, processing, research and development, testing, packaging, and storage, etc. These activities demand precise temperature and humidity control depending on the nature of raw materials/tools used, safety of the equipment, and thermal comfort of the occupants. Desiccant dehumidifiers provide energy-efficient, cost-effective, low maintenance solutions, and offer application-specific climate control strategies [60]. The following section highlights their role in the pharmaceutical and electronics manufacturing industries.

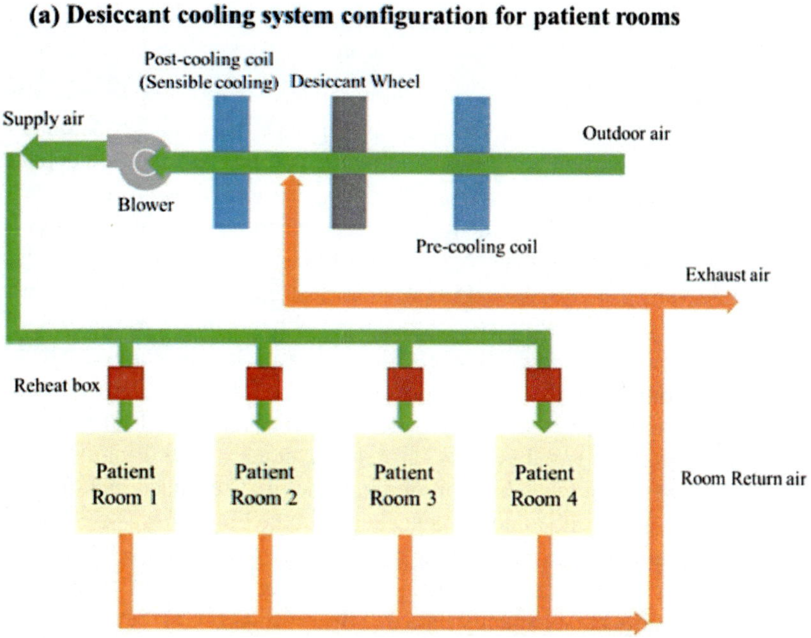


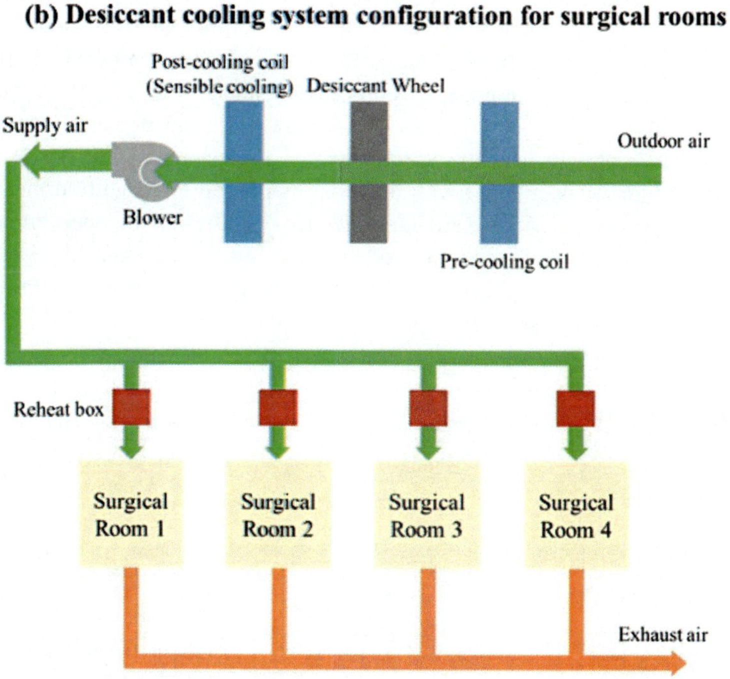


**Fig. 5.15** Desiccant cooling system configurations for **a** patient rooms and **b** surgical rooms for healthcare sectors. For patient rooms, a portion of the return air from the rooms is mixed with the fresh air intake and then supplied while 100% outdoor air is used for surgical care suites

### 5.4.1 Pharmaceutical Manufacturing

Pharmaceutical manufacturing is highly sensitive to moisture levels, and the temperature set points are not critical since the effects of thermal expansion or contraction are minimal. Consequently, maintaining the building temperature in pharmaceutical production is mainly governed by the occupant thermal comfort [61]. While temperature levels are not critical, achieving a closed control of the humidity at 10–15% RH is a key factor as it eliminates 90% of moisture absorption [62]. This leads to excellent manufacturing standards and improvement in the shelf life of tablets and capsules. Table 5.2 lists the quality issues associated with moisture exposure in different steps of pharmaceutical manufacturing [63]. The tablet/capsule production mainly involves mixing, compression, and packaging operations. The raw ingredients (usually dry powders) are ground into fine powders and mixed in industrial blenders that rotate at high speeds. If wet mixing is carried out, an additional drying step is required to ensure that the powders are appropriately dried. The mixed and dried powders are then compressed into tablets and packaged in the blister form or pellets. When there is any significant deviation from the humidity setpoint, the tablets/capsules stick together or break apart upon exposure to higher moisture content. This affects the product quality and is directly linked to the revenue generated by the pharmaceutical company [64].

Conventional MVC air-conditioning systems produce air at highly saturated conditions that significantly deviate from the humidity requirements of pharmaceutical companies. Further, condensate formation and water leakage in the ductwork encourage microbiological reproduction and affect indoor air quality [65]. As a result, an energy-efficient low dew point desiccant dehumidifier is normally preferred to supply the dry air requirements of pharmaceutical industries. The desiccant dehumidifier is coupled with the existing central air-conditioning facility, and the two cooling technologies are integrated to yield the desired operating conditions. Figure 5.16 illustrates a schematic of a desiccant wheel enhanced air-conditioning system employed for a pharmaceutical manufacturing facility [66]. In this system, the fresh air (around 20% of the total supply air) is dehumidified and pre-cooled via

**Table 5.2** Demerits of moisture exposure on pharmaceutical manufacturing processes

| Manufacturing processes | Demerits of moisture exposure |
|---|---|
| Mixing | Challenges in blending. Leads to lump formation |
| Coating | Uneven coating and increase in roughness |
| Tableting | Ineffective compression and crack formation |
| Blister packaging | Tablets sticking together and affecting shelf life |
| Capsule compaction | Changes in product consistency and discoloration |

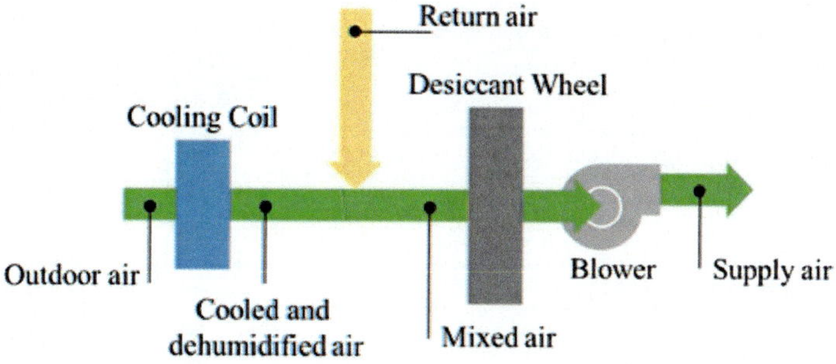


**Fig. 5.16** A typical desiccant wheel enhanced air-conditioning system for a pharmaceutical manufacturing facility

an MVC chiller system. This air, which is at near saturation levels, is then mixed with air returning from the room. The mixed air then passes through a desiccant wheel dehumidifier, where the air is dehumidified to supply air at 5% RH. The pre-cooling step via MVC technology is highly cost-effective since there is more potential for cooling and dehumidification to occur at high temperature and humidity environments. In addition, since the air passing through the desiccant wheel is pre-cooled, exothermic sorption heat does not overheat the supply air while providing extremely low humidity specifications.

## *5.4.2 Electronics Manufacturing*

Desiccant dehumidifiers can operate as specialized standalone units in the different stages of the electronics manufacturing industry, namely, production, assembly, processing, storage, and operations. It is estimated that over 25% of defects in electronics equipment are attributed to excessive moisture exposure during the manufacturing stage. Since moisture affects the performance critically at every stage, different processes demand a unique set of temperature and humidity conditions. For instance, fabricating lithium batteries call for extremely low humidity conditions (around less than 1% RH). This is because lithium oxidizes rapidly in the presence of moisture and explodes [67].

In another case, when the electronic components are rapidly heated and cooled, condensation occurs, leading to ineffective calibration that affects performance [68]. Additionally, in the front-line production of semiconductors, photo-sensitive polymers called photoresists are used to selectively protect the devices during the etching process. The photoresists are hygroscopic polymers that readily absorb moisture and cause circuit line bridging. The room RH needs to be regulated between 20 and 35% at a low dew point temperature of −4 °C to +5 °C to avoid any impending failure due to excessive moisture presence.

**Table 5.3** Recommended temperature and humidity limits for selected electronic production processes [71]

| Electronics manufacturing processes | Temperature (°C) | RH (%) |
|---|---|---|
| Thermostat and humidistat assembly and calibration | 24 | 50–55 |
| Capacitor winding | 22.5 | 50 |
| Rectifier processing | 23 | 30–40 |
| Circuit breakers assembly and testing | 24 | 30–60 |
| Lithium battery manufacturing | 22 | 0–1 |
| Metering devices assembly and test | 24 | 60–63 |
| Transformer winding | 22.5 | 15 |
| Instrument manufacturing laboratory | 20–22 | 50–55 |
| Lightning arrestor assembly | 20 | 20–40 |

During the assembly stage of printed circuit boards (PCBs), any exposure to high moisture negatively affects the performance. Moisture increases their electrical resistance, reduces capacitance, promotes micro-level corrosion, and ultimately leads to surface defects and adhesion failures. On the other hand, when the air is too dry, electrostatic discharge (ESD) occurs due to the sudden electricity flow between the charged objects. This phenomenon is undesired as it damages the internal circuitry of PCBs. Dry air promotes the creation of static charges, whereas humid air assists in dissipating them. Therefore, maintaining strict moisture control is necessary for proper device functioning [69, 70]. Table 5.3 lists the standard operating temperature and humidity requirements for different electronics manufacturing processes [71].

## 5.5 Conclusions

In this chapter, several potential applications of solid desiccant dehumidifiers have been comprehensively reviewed. In standalone heat pumps, evaporators and condensers can be replaced with DCHEs, and an energy-efficient strategy of simultaneously handling both sensible and latent cooling loads can be obtained by adopting to more advanced desiccants. In addition, DCHEs are found to be suitable for closed-cycle adsorption chilling applications. Specifically, MOFs with their excellent structural flexibility, S-shaped isotherms, high cyclic stability, and adsorption heat equivalent to the latent heat of vaporization of water demonstrate extraordinary potential in chillers. The lower regeneration temperature requirement enhances the technology's suitability for integration with sustainable energy sources such as solar/waste heat.

Apart from heat transformation applications, desiccant dehumidifiers have also been employed for thermochemical storage systems, harvesting potable water from the atmosphere, and precise moisture control. The thermochemical storage systems are more efficient and less bulky than sensible and latent heat storage devices since the energy can be stored as the chemical potential of desiccants without any significant loss. More importantly, the energy density of the desiccant is of the range spanning 200–500 $kWh/m^3$ as opposed to 50 $kWh/m^3$ for sensible and 100 $kWh/m^3$ for latent heat storage systems. As atmospheric water harvesters, next-generation desiccants present a valuable proposition due to their advancements in sorption-desorption characteristics. Lastly, desiccant dehumidifiers are widely employed to achieve effective localized temperature and humidity moisture control in different functions of building cooling systems and industrial air-conditioners. Thermal comfort is a fundamental necessity to realize both excellent living and working environment since humidity levels lower than 40% RH cause excessive dryness and promote bacterial/viral/mold growth when RH exceeds 80%. Likewise, attaining specified moisture control leads to excellent manufacturing standards, extends the shelf life of equipment and tools, and directly relates to higher profits.

## References

1. Tu YD, Wang RZ, Ge TS, Zheng X. Comfortable, high-efficiency heat pump with desiccant-coated, water-sorbing heat exchangers 2017:1–10. https://doi.org/10.1038/srep40437.
2. Enteria N, Mizutani K, Monma Y, Akisaka T, Okazaki N. Experimental evaluation of the new solid desiccant heat pump system in Asia-Pacific climatic conditions. Appl Therm Eng. 2011;31:243–57. https://doi.org/10.1016/j.applthermaleng.2010.09.004.
3. Zhang L, Hashimoto K, Hasegawa H, Saikawa M. Performance analysis of a heat pump system with integrated desiccant for electric vehicles. Int J Refrig. 2018;86:154–62. https://doi.org/10.1016/j.ijrefrig.2017.10.036.
4. Yang TY, Ge TS, Lu FL, Dai YJ, Wang RZ. A novel semi-coupled solid desiccant heat pump system—Part 1: simulation study. Int J Refrig. 2020;120:150–60. https://doi.org/10.1016/j.ijrefrig.2020.09.005.
5. Ge TS, Yang TY, Lu FL, Dai YJ, Wang RZ. A novel semi-coupled solid desiccant heat pump system Part 2: experimental investigation. Int J Refrig. 2021;121:86–94. https://doi.org/10.1016/j.ijrefrig.2020.10.008.
6. Hua LJ, Ge TS, Wang RZ. A mathematical model to predict the performance of desiccant coated evaporators and condensers. Int J Refrig 2020. https://doi.org/10.1016/j.ijrefrig.2019.10.001.
7. Hua LJ, Jiang Y, Ge TS, Wang RZ. Experimental investigation on a novel heat pump system based on desiccant coated heat exchangers. Energy. 2017;142:96–107. https://doi.org/10.1016/j.energy.2017.07.141.
8. Hua LJ, Ge TS, Wang RZ. Extremely high efficient heat pump with desiccant coated evaporator and condenser. Energy. 2019;170:569–79. https://doi.org/10.1016/j.energy.2018.12.169.
9. Kummer H, Jeremias F, Warlo A, Füldner G, Fröhlich D, Janiak C, et al. A functional full-scale heat exchanger coated with aluminum fumarate metal-organic framework for adsorption heat transformation. Ind Eng Chem Res. 2017;56:8393–8. https://doi.org/10.1021/acs.iecr.7b00106.
10. Cui S, Qin M, Marandi A, Steggles V, Wang S, Feng X, et al. Metal-Organic Frameworks as advanced moisture sorbents for energy-efficient high temperature cooling. Sci Rep. 2018;8:2–10. https://doi.org/10.1038/s41598-018-33704-4.
11. Prabakaran V. Studying the transport phenomena of superabsorbents: an experimental and theoretical approach 2020.
12. Vivekh P, Kumja M, Bui DT, Chua KJ. Recent developments in solid desiccant coated heat exchangers—a review. Appl Energy. 2018;229:778–803. https://doi.org/10.1016/j.apenergy.2018.08.041.

13. Kim DS, Chang YS, Lee DY. Modelling of an adsorption chiller with adsorbent-coated heat exchangers: feasibility of a polymer-water adsorption chiller. Energy. 2018;164:1044–61. https://doi.org/10.1016/j.energy.2018.09.041.
14. Ugale VD, Pitale AD. A review on working pair used in adsorption cooling system. Int J Air-Conditioning Refrig. 2015;23:1–11. https://doi.org/10.1142/S2010132515300013.
15. Mitsubishi Plastics to Test-Market Easy-to-Install Compact Adsorption Chiller with Integrated Cooling Tower to the U.S. Market | Business Wire n.d. https://www.businesswire.com/news/home/20120117006242/en/Mitsubishi-Plastics-Test-Market-Easy-to-Install-Compact-Adsorption-Chiller. Accessed April 4, 2018.
16. AQSOA honeycomb wheel and AQSOA coated heat exchanger n.d. http://www.aaasaveenergy.com/products/001/pdf/AQSOA_1210E.pdf. Accessed April 4, 2018.
17. Okamoto K, Teduka M, Nakano T, Kubokawa S, Kakiuchi H. The Development of Aqsoa water vapor adsorbent and Aqsoa coated heat exchanger. Proc Int Symp Innov Mater Process Energy Syst. 2010, Singapore: Research Publishing Services; 2010, p. 27–32. https://doi.org/10.3850/978-981-08-7614-2_IMPRES043.
18. Aristov YI. Current progress in adsorption technologies for low-energy buildings. Futur Cities Environ. 2015;1:10. https://doi.org/10.1186/s40984-015-0011-x.
19. Kayal S, Baichuan S, Saha BB. Adsorption characteristics of AQSOA zeolites and water for adsorption chillers. Int J Heat Mass Transf. 2016;92:1120–7. https://doi.org/10.1016/J.IJHEATMASSTRANSFER.2015.09.060.
20. Goldsworthy MJ. Measurements of water vapour sorption isotherms for RD silica gel, AQSOA-Z01, AQSOA-Z02, AQSOA-Z05 and CECA zeolite 3A. Microporous Mesoporous Mater. 2014;196:59–67. https://doi.org/10.1016/j.micromeso.2014.04.046.
21. Tannert N, Ernst SJ, Jansen C, Bart HJ, Henninger SK, Janiak C. Evaluation of the highly stable metal-organic framework MIL-53(Al)-TDC (TDC = 2,5-thiophenedicarboxylate) as a new and promising adsorbent for heat transformation applications. J Mater Chem A. 2018;6:17706–12. https://doi.org/10.1039/c8ta04407d.
22. Uma K, Pan GT, Yang TCK. The preparation of porous sol-gel silica with metal organic framework MIL-101(Cr) by microwave-assisted hydrothermal method for adsorption chillers. Materials (Basel). 2017;10:610. https://doi.org/10.3390/ma10060610.
23. Xia X, Cao M, Liu Z, Li W, Li S. Elucidation of adsorption cooling characteristics of Zr-MOFs: effects of structure property and working fluids. Chem Eng Sci. 2019;204:48–58. https://doi.org/10.1016/j.ces.2019.04.006.
24. Lenzen D, Bendix P, Reinsch H, Fröhlich D, Kummer H, Möllers M, et al. Scalable green synthesis and full-scale test of the metal-organic framework CAU-10-H for use in adsorption-driven chillers. Adv Mater. 2018;30:1705869. https://doi.org/10.1002/adma.201705869.
25. Hanikel N, Prévot MS, Fathieh F, Kapustin EA, Lyu H, Wang H, et al. Rapid cycling and exceptional yield in a metal-organic framework water harvester. ACS Cent Sci 2019. https://doi.org/10.1021/acscentsci.9b00745.
26. Zheng G, Chen M, Yin J, Zhang H, Liang X, Zhang J. Metal organic frameworks derived nano materials for energy storage application. Int J Electrochem Sci. 2019;14:2345–62. https://doi.org/10.20964/2019.03.28.
27. Xia W, Mahmood A, Zou R, Xu Q. Metal-organic frameworks and their derived nanostructures for electrochemical energy storage and conversion. Energy Environ Sci. 2015;8:1837–66. https://doi.org/10.1039/c5ee00762c.
28. Xu G, Nie P, Dou H, Ding B, Li L, Zhang X. Exploring metal organic frameworks for energy storage in batteries and supercapacitors. Mater Today. 2017;20:191–209. https://doi.org/10.1016/j.mattod.2016.10.003.
29. Wu H Bin, Lou XW. Metal-organic frameworks and their derived materials for electrochemical energy storage and conversion: Promises and challenges. Sci Adv 2017;3:1–17. https://doi.org/10.1126/sciadv.aap9252.
30. Lefebvre D, Tezel FH. A review of energy storage technologies with a focus on adsorption thermal energy storage processes for heating applications. Renew Sustain Energy Rev. 2017;67:116–25. https://doi.org/10.1016/J.RSER.2016.08.019.

31. Michel B, Mazet N, Mauran S, Stitou D, Xu J. Thermochemical process for seasonal storage of solar energy: characterization and modeling of a high density reactive bed. Energy. 2012;47:553–63. https://doi.org/10.1016/J.ENERGY.2012.09.029.
32. Thu K, Nasruddin N, Mitra S, Saha BB. Sorption-based energy storage systems: a review. Makara J Technol. 2019;23:16. https://doi.org/10.7454/mst.v23i1.3691.
33. Hauer A. Evaluation of adsorbent materials for heat pump and thermal energy storage applications in open systems. Adsorption. 2007;13:399–405. https://doi.org/10.1007/s10450-007-9054-0.
34. Finck C, Henquet E, van Soest C, Oversloot H, de Jong A-J, Cuypers R, et al. Experimental Results of a 3 kWh thermochemical heat storage module for space heating application. Energy Procedia. 2014;48:320–6. https://doi.org/10.1016/J.EGYPRO.2014.02.037.
35. Studies of the water adsorption on Zeolites and modified mesoporous materials for seasonal storage of solar heat. Sol Energy 2004;76:339–44. https://doi.org/10.1016/J.SOLENER.2003.07.036.
36. Jarimi H, Aydin D, Yanan Z, Ozankaya G, Chen X, Riffat S. Review on the recent progress of thermochemical materials and processes for solar thermal energy storage and industrial waste heat recovery. Int J Low-Carbon Technol. 2019;14:44–69. https://doi.org/10.1093/ijlct/cty052.
37. Elsayed A, Elsayed E, AL-Dadah R, Mahmoud S, Elshaer A, Kaialy W. Thermal energy storage using metal–organic framework materials. Appl Energy 2017;186:509–19. https://doi.org/10.1016/j.apenergy.2016.03.113.
38. Garzón-Tovar L, Pérez-Carvajal J, Imaz I, Maspoch D. Composite salt in porous metal-organic frameworks for adsorption heat transformation. Adv Funct Mater. 2017;27:1606424. https://doi.org/10.1002/adfm.201606424.
39. HEATS Program Overview B. PROGRAM OVERVIEW. n.d.
40. Fathieh F, Kalmutzki MJ, Kapustin EA, Waller PJ, Yang J, Yaghi OM. Practical water production from desert air. Sci Adv. 2018;4:1–10. https://doi.org/10.1126/sciadv.aat3198.
41. Kalmutzki MJ, Diercks CS, Yaghi OM. Metal-organic frameworks for water harvesting from air. Adv Mater. 2018;30:1–26. https://doi.org/10.1002/adma.201704304.
42. Kim H, Yang S, Rao SR, Narayanan S, Kapustin EA, Furukawa H, et al. Water harvesting from air with metal-organic frameworks powered by natural sunlight n.d.
43. Bui DT, J. CK, Gordon JM. Water harvesting from air with metal-organic frameworks powerd by natural sunlight. Science (80-) 2017;356:430–4.
44. Kim H, Rao SR, Narayanan S, Kapustin EA, Yang S, Furukawa H, et al. Response to comment on " Water harvesting from air with metal- organic frameworks powered by natural sunlight" 2017;3139:1–4.
45. Wang W, Xie S, Pan Q, Dai Y, Wang R, Ge T. Air-cooled adsorption-based device for harvesting water from island air. Renew Sustain Energy Rev. 2021;141: https://doi.org/10.1016/j.rser.2021.110802.
46. Turner SC, Paliaga G, Lynch BM, Arens EA, Aynsley RM, Brager GS, et al. Standard 55-2010 Thermal environmental conditions for human occupancy, Atlanta. GA. ASHRAE. 2010;2010:30.
47. Moisture control and ventilation—WHO Guidelines for Indoor Air Quality—NCBI Bookshelf n.d. https://www.ncbi.nlm.nih.gov/books/NBK143947/. Accessed May 5, 2021.
48. Shehadi M. Review of humidity control technologies in buildings. J Build Eng. 2018;19:539–51. https://doi.org/10.1016/j.jobe.2018.06.009.
49. Alezi D, Shkurenko A, Samin U, Aggarwal H, Weseliński ŁJ, Belmabkhout Y, et al. Reticular chemistry in action: a hydrolytically stable MOF capturing twice its weight in adsorbed water. Chem. 2018;4:94–105. https://doi.org/10.1016/j.chempr.2017.11.005.
50. ComDry M210X Desiccant Dehumidifier | Munters n.d. https://www.munters.com/en/munters/products/dehumidifiers/comdry-m210x-desiccant-dehumidifier/. Accessed May 6, 2021.
51. DSS Desiccant System Solution™ | Munters n.d. https://www.munters.com/en/munters/products/dehumidifiers/dss-desiccant-system-solution–ef3fc1ed/. Accessed May 6, 2021.
52. How humidity control can reduce viral infectivity and longevity Executive summary 2020.

53. COVID Live Update: 152,491,572 Cases and 3,199,892 Deaths from the Coronavirus—Worldometer n.d. https://www.worldometers.info/coronavirus/. Accessed May 2, 2021.
54. Advances in desiccant-based dehumidification. Trane. 2005;34:1–6.
55. Coronavirus SARS-CoV-2 spreads more indoors at low humidity: Indian-German research team recommends at least 40 percent humidity in public buildings—ScienceDaily n.d. https://www.sciencedaily.com/releases/2020/08/200820102503.htm. Accessed May 2, 2021.
56. Ahlawat A, Wiedensohler A, Mishra SK. An overview on the role of relative humidity in airborne transmission of sars-cov-2 in indoor environments. Aerosol Air Qual Res. 2020;20:1856–61. https://doi.org/10.4209/aaqr.2020.06.0302.
57. Cui X, Mohan B, Islam MR, Chua KJ. Investigating the energy performance of an air treatment incorporated cooling system for hot and humid climate. Energy Build. 2017;151:217–27. https://doi.org/10.1016/j.enbuild.2017.06.059.
58. Healthcare Facility, Humidity Control in Hospitals Industry | Bry-Air n.d. https://www.bryair.com/en/uae-en/industries-applications/applications-of-dehumidifiers-and-dryers/production-and-processing/healthcare-facility/. Accessed April 29, 2021.
59. Hospital Dehumidification Systems—Dehumidification for Medical Facilities | Munters n.d. https://www.munters.com/en/industries/healthcare/. Aaccessed April 29, 2021.
60. Application of Dehumidifiers for Various Industries—Industry Today n.d. https://industrytoday.com/dehumidifier-applications/. Accessed May 6, 2021.
61. Perfect climate for pharmaceutical manufacture | Munters n.d. https://www.munters.com/en/munters/cases/pharmaceutical-perfect-climate/. Accessed May 1, 2021.
62. Pharmaceutical Industry—Destech.eu—The most Advanced Desiccant Technologies n.d. https://destech.eu/applications/production/pharmaceutical-industry/. Accessed May 1, 2021.
63. Quality Air For The Pharmaceutical Industry | Bryair n.d. https://www.bryair.com/technical-articles/quality-air-for-the-pharmaceutical-industry/. Accessed May 1, 2021.
64. Dehumidifiers for pharmaceutical and cosmetics industries | Fisair n.d. https://fisair.com/dehumidifiers-for-pharmaceutical-and-cosmetics-industries/. Accessed May 1, 2021.
65. Pharmaceutical Product & Laboratory Dehumidification System | Munters n.d. https://www.munters.com/en/industries/pharmaceutical/. Accessed May 1, 2021.
66. Harriman LG. The dehumidification handbook. 2002.
67. Electronic Manufacturing Dehumidifiers—Dehumidification for Electronic Assembly | Munters n.d. https://www.munters.com/en/industries/electronics/. Accessed April 30, 2021.
68. Environmental Control Systems for Dehumidification Applications In the Electronics Industry—Bry Air n.d. http://bry-air.com/casestudies/dehumidification-environmental-control-systems-for-electronics-industry/. Accessed April 30, 2021.
69. Humidity For Electronics Manufacturing | Condair n.d. https://www.condair.com/humidifiers-for-electronics. Accessed April 30, 2021.
70. Why humidity control matters in electronics n.d. https://airtecsolutions.com/applications/electronics. Accessed April 30, 2021.
71. Dehumidification Application PRINTED CIRCUIT BOARD ASSEMBLY. n.d.